Springer Series in
MATERIALS SCIENCE
40

Springer-Verlag Berlin Heidelberg GmbH

Physics and Astronomy

ONLINE LIBRARY

http://www.springer.de/phys/

Springer Series in
MATERIALS SCIENCE

Editors: R. Hull   R. M. Osgood, Jr.   H. Sakaki   A. Zunger

The Springer Series in Materials Science covers the complete spectrum of materials physics, including fundamental principles, physical properties, materials theory and design. Recognizing the increasing importance of materials science in future device technologies, the book titles in this series reflect the state-of-the-art in understanding and controlling the structure and properties of all important classes of materials.

Series homepage – http://www.springer.de/phys/books/ssms/

Volumes 1–25 are listed at the end of the book.

A. Zschunke (Ed.)

Reference Materials in Analytical Chemistry

A Guide for Selection and Use

With 30 Figures and 34 Tables

 Springer

Prof. Dr. Adolf Zschunke
Bundesanstalt für Metallforschung und -prüfung
Richard-Willstätter-Strasse 11,
12489 Berlin, Germany

Series Editors:

Prof. Alex Zunger
NREL
National Renewable Energy Laboratory
1617 Cole Boulevard
Golden Colorado 80401-3393, USA

Prof. R. M. Osgood, Jr.
Microelectronics Science Laboratory
Department of Electrical Engineering
Columbia University
Seeley W. Mudd Building
New York, NY 10027, USA

Prof. Robert Hull
University of Virginia
Dept. of Materials Science and Engineering
Thornton Hall
Charlottesville, VA 22903-2442, USA

Prof. H. Sakaki
Institute of Industrial Science
University of Tokyo
7-22-1 Roppongi, Minato-ku
Tokyo 106, Japan

ISSN 0933-033x
ISBN 978-3-642-63097-2   ISBN 978-3-642-56986-9 (eBook)
DOI 10.1007/978-3-642-56986-9

Library of Congress Cataloging-in-Publication Data.

Reference materials in analytical chemistry: a guide for selection and use/A. Zschunke (ed.). p.cm.– (Springer series in materials science, ISSN 0933-033x; 40). Includes bibliographical references and index. 1. Chemistry, analytic–Quality control. 2. Materials–Standards. I. Zschunke, A. (Adolf), 1937- . II. Springer series in materials science; v. 40. QD75.4.Q34 r43 2000 543'.00218–dc21 00-039464

© Springer-Verlag Berlin Heidelberg 2000

Originally published by Springer-Verlag Berlin Heidelberg New York in 2000.
Softcover reprint of the hardcover 1st edition 2000

Typesetting: Data conversion by Steingraeber Satztechnik GmbH, Heidelberg
Cover concept: eStudio Calamar Steinen
Cover production: *design & production* GmbH, Heidelberg

Printed on acid-free paper   SPIN: 10748139   57/3144/tr   5 4 3 2 1 0

Preface

Analytical chemistry is increasingly becoming an important basis for decision making in science, economy, trade, health care, environmental and consumer protection, sports and jurisdiction. In all spheres, analytical results need to fulfill a purpose. This means being accurate to a specified extent according to the required level of uncertainty, as well as being comparable, depending on the range of validity. In many cases, this range is global. Reference materials are important tools for meeting these demands. Driving forces behind the increasing diversification of reference materials also include the rapid development of analytical chemistry, the development of new methods and applications, and the inclusion of new analytes, matrices and materials. The growing need for reference materials is giving rise to an increasing demand for information, orientation and stimulation. Laboratory staff have questions on calibration, validation, home-made reference materials and certified reference materials. This book is a handbook for analytical chemists and technical personnel in analytical laboratories, where it should become a standard reference work.

Berlin, February 2000   *A. Zschunke*

Contents

5   Reference Materials in Environmental Studies
Irene Nehls and Tin Win

6   Reference Materials in Clinical and Forensic Toxicological Analysis
Fritz Pragst and Wolf-Rüdiger Külpmann

7   Use of Reference Materials in Gas Analysis
Bruno Reimann

8   The International Network
Harry Klich

List of Abbreviations

AAS	Atomic absorption spectrometry
AOAC	Association of Official Analytical Chemists
AQCS	Analytical Quality Control Services (IAEA)
BAM	Bundesanstalt für Materialforschung und -prüfung (Germany)
BAS	Bureau of Analysed Samples (UK)
BCR	Bureau Communautaire de Réference (EC bureau of reference)
BCS	British Chemical Standards
BERM	International symposium on Biological and Environmental Reference Materials
BIPM	International Bureau of Weights and Measures (Paris)
BMEMC	Beijing Municipal Environmental Monitoring Centre
CAP	College of American Pathologists
CBNM	Central Bureau for Nuclear Measurements (EC)
CCQM	Comité Consultatif pour la Quantité de Matière
CEN	Comité Européen de Normalisation
CENAM	Centro Nacional de Metrologia (Mexico)
CERM	Central European Conference on Reference Materials
CIPM	International Committee for Weights and Measures
COCRI	Cereal and Oil Chemistry Research Institute (China)
COMAR	Database for certified reference materials
COWS (WASP)	Commission on World Standards of the World Association of Societies of Pathology
CRM	Certified reference material
CSTL	NIST Chemical Science and Technology Laboratory
CTIF	Centre Technique des Industries de la Fonderie (France)
DAD	Photodiode array detection
DANREF	Danish center for chemical reference materials
DSC	Differential scanning calorimetry
DUREM	Indian national workshop on development and use of environmental reference materials
EA	European co-operation for Accreditation
EC	European Community

ECCLS	European Committee for Clinical Laboratory Standards
ECISS	European Committee for Iron and Steel Standardization
EIGA	European Industrial Gases Association
EMCHJ	Enivronmental Monitoring Centre in Heilong Jiang (China)
EMPA	Eidgenössische Materialprüfungs- und Forschungs- anstalt (CH)
EP	European Pharmacopoeia
EURACHEM	Association of European Chemical Laboratories
EUROLAB	European Federation of National Associations of Measurement, Testing and Analytical Laboratories
FDA	Federal Drug Administration (USA)
FDSI	Food Detection Science Institute (China)
FECS	Federation of European Chemical Societies
FIA	Flow injection analysis
GC	Gas chromatography
GD	Glow discharge
GDMB	Gesellschaft für Bergbau, Metallurgie, Rohstoff und Umwelttechnik (Germany)
GTFch	Gesellschaft für Forensische und Toxikologische Chemie (Germany)
GUM	Guide to the Expression of Uncertainty in Measurement
HOMEX	BAM programme for data evaluation
HPLC	High performance liquid chromatography
IAEA	International Atomic Energy Agency
ICP	Inductively coupled plasma
IDMS	Isotope dilution mass spectrometry
IEC	International Electrotechnical Commission
IEC	Institute of Environmental Chemistry (China)
IFCC	International Federation of Clinical Chemistry
IGGE	Institute of Geophysical and Geochemical Exploration (China)
ILAC	International Laboratory Accreditation Cooperation
IMEP	International Measurement Evaluation Programme
INAA	Instrumental neutron activation analysis
IOC	International Olympic Committee
IRMM	Institute for Reference Materials and Measurements (CEC Joint Research Centre)
IRSID	Institut de Recherches de la Sidérurgie Française (France)
ISO	International Standardization Organization
ITS-90	International Temperature Scale of 1990
IUPAC	International Union of Pure and Applied Chemistry
JCS	Japan Ceramic Society

JISF	Japanese Iron and Steel Federation
LA	Laser ablation
LGC	Laboratory of the Government Chemist (UK)
LNE	Laboratoire National d'Essais (France)
MDA	methylenedioxyamphetamine
MPI	Max Planck Institute
MS	Mass spectrometry
NAA	Neutron activation analysis
NBS	National Bureau of Standards (US) (since 1980 NIST)
NC-CRM	Chinese national research centre for certified reference materials
NCRMWG	Nordic Certified Reference Material Working Group
NIES	National Institute for Environmental Studies (Japan)
NIST	National Institute of Standards and Technology (US)
NMi	Nederlands Meetinstitut (Netherlands Measurements Institute)
NMR	Nuclear magnetic resonance
NPL	National Physical Laboratory (UK)
NRC	National Research Council, Ottawa (Canada)
NTRM^{TM}	NIST traceable reference material
OES	Optical emission spectrometry
OIML	International Organization of Legal Metrology
PAH	Polycyclic aromatic hydrocarbon
PCB	Polychlorinated biphenyl
PPRI	Paper and Pulp Research Institute, Bratislava (Slovakia)
PRM	Pharmaceutical reference materials
PT	Proficiency testing
PTB	Physikalisch-Technische Bundesanstalt (Germany)
PTFE	Polytetrafluoroethylene
QA	Quality assurance
QC	Quality control
REMTAF	Reference Material Task Force of India
RIG	Research Institute of Geology (China)
RM	Reference material
SD	Spark discharge
SI	International system of units
SIM	Slovak Institute of Metrology
SIMS	Selected ion mass spectrometry
SINR	Shanghai Institute of Nuclear Research (China)
SITT	Shanghai Institute of Testing Technology (China)
SM&T	Standards, Measurement and Testing programmes (EC)
SOFT	Society of Forensic Toxicologists
SRM^{R}	NIST standard reference material
STA	Systematic toxicological analysis

UNEP-HEM	United Nations Environmental Programme, Harmonization of Environmental Measurement
USP	US Pharmacopoeia
VAMAS	Versailles Project on Advanced Materials and Standards
VDEh	Verein Deutscher Eisenhüttenleute (Germany)
VIM	International vocabulary of basic and general terms
WAU	Wageningen Agriculture University
WHO	World Health Organization
XRF	X-ray fluorescence analysis

List of Contributors

C. Beck
National Institute of Standards
and Technology (NIST)
Analytical Chemistry Division
Gaithersburg, MD 20899-8390
USA

J. Colbert
National Institute of Standards
and Technology (NIST)
Standard Reference Materials Program
Gaithersburg, MD 20899-2000
USA

J. Fassett
National Institute of Standards
and Technology (NIST)
Analytical Chemistry Division
Gaithersburg, MD 20899-8390
USA

R. Gettings
National Institute of Standards
and Technology (NIST)
Standard Reference Materials Program
Gaithersburg, MD 20899-2000
USA

T. Gills
National Institute of Standards
and Technology (NIST)
Standard Reference Materials Program
Gaithersburg, MD 20899-2000
USA

R. Greenberg
National Institute of Standards
and Technology (NIST)
Analytical Chemistry Division
Gaithersburg, MD 20899-8390
USA

F. Guenther
National Institute of Standards
and Technology (NIST)
Analytical Chemistry Division
Gaithersburg, MD 20899-8390
USA

Werner Hässelbarth
Bundesanstalt für Materialforschung
und -prüfung
Richard-Willstätter-Strasse 11
12489 Berlin
Germany

Harry Klich
Bundesanstalt für Materialforschung
und -prüfung
Richard-Willstätter-Strasse 11
12489 Berlin
Germany

G. Kramer
National Institute of Standards
and Technology (NIST)
Analytical Chemistry Division
Gaithersburg, MD 20899-8390
USA

Wolf-Rüdiger Külpmann
Medizinische Hochschule Hannover
Institut für Klinische Chemie I
Carl-Neuberg-Strasse 1
30625 Hannover
Germany

Siegfried Noack
Bundesanstalt für Materialforschung
und -prüfung
Richard-Willstätter-Strasse 11
12489 Berlin
Germany

B. MacDonald
National Institute of Standards
and Technology (NIST)
Standard Reference Materials Program
Gaithersburg, MD 20899-2000
USA

R. Parris
National Institute of Standards
and Technology (NIST)
Analytical Chemistry Division
Gaithersburg, MD 20899-8390
USA

Ralf Matschat
Bundesanstalt für Materialforschung
und -prüfung
Richard-Willstätter-Strasse 11
12489 Berlin
Germany

Fritz Pragst
Institut für Rechtsmedizin der
Humboldt-Universität
Hannoversche Strasse 6
10115 Berlin
Germany

W. May
National Institute of Standards
and Technology (NIST)
Analytical Chemistry Division
Gaithersburg, MD 20899-8390
USA

Bruno Reimann
Messer Griesheim GmbH
Postfach 10 09 62
47009 Duisburg
Germany

Klaus Meyer
Bundesanstalt für Materialforschung
und -prüfung
Richard-Willstätter-Strasse 11
12489 Berlin
Germany

Tomas Tamberg
Bundesanstalt für Materialforschung
und -prüfung
Richard-Willstätter-Strasse 11
12489 Berlin
Germany

Irene Nehls
Bundesanstalt für Materialforschung
und -prüfung
Richard-Willstätter-Strasse 11
12489 Berlin
Germany

Tin Win
Bundesanstalt für Materialforschung
und -prüfung
Richard-Willstätter-Strasse 11
12489 Berlin
Germany

S. Wise
National Institute of Standards
and Technology (NIST)
Analytical Chemistry Division
Gaithersburg, MD 20899-8390
USA

Adolf Zschunke
Bundesanstalt für Materialforschung
und -prüfung
Richard-Willstätter-Strasse 11
12489 Berlin
Germany

1 Introduction

Adolf Zschunke

Reference samples and reference materials have been used in analytical chemistry ever since analytical chemistry first existed. The necessity to compare analytical results between laboratories and between countries was already felt in 1906, when the US National Bureau of Standards (NBS) initiated a program to provide reference materials, originally known as standard samples, largely in response to the needs of the metal industry. The reliability of all analytical results is completely dependent on the availability of reference materials and nowadays nearly all branches of analytical chemistry declare an urgent need for reference materials.

This book was prepared with the objective of improving the understanding of the fundamentals of reference materials use. The basic concept is to consider metrology as a powerful tool in analytical chemistry and not analytical chemistry as a part of metrology. Analytical chemistry is not equivalent to chemical measurement. But chemical measurement is a crucial part of analytical chemistry, emphasizing the measuring aspects. Chemical measurements are also applied in other scientific disciplines such as biology, physics, medicine and many other testing fields. Reference materials have to reflect the complete extent of analytical chemistry and chemical measurements. They have also to reflect the developments of analytical chemistry, for example, process analysis.

The main focus of the book is analytical chemistry, but there is no sharp separation from biological, medical or physical analysis. The book enables analytical chemists to judge the different kinds of reference materials and their usefulness.

The book is arranged by sections in a logical progression, starting with the basic concepts of analytical chemistry and the need for standards and quality assurance, followed by discussions of classification and certification of reference materials, of the use of reference materials in various fields of analytical chemistry and of international normative documents in those fields. The treatment of each subject in a section is not necessarily exhaustive and only selected examples of the use of reference materials are presented.

1.1 Analytical Thinking

Analytical chemistry is not an independent science. As a scientific discipline it is related to other disciplines of chemistry as well as to physics, biology, medicine and other sciences. The term analytical chemistry includes the two terms analysis and chemistry.

Analysis is a way of thinking and is elemental to science. It means:

- structuring of a whole into its parts (each problem has to be divided into as many parts as needed for its solution);
- defining types, properties and relations of the parts;
- acceptance of the validity of logical rules;
- tracing things back to elements, reasons, driving forces, terms and principles;
- introduction of terms of reference, reference frames, units of measurement, standards and scales;
- aiming at completeness of statements (avoid missing something);
- thought has to follow a course from simple to complex;
- application of selection criteria;
- definition of targets and strategies;
- establishment of hypotheses, laws, standards and rules.

The use of analytical methods marked the beginning of modern science. Analytical chemistry is the materialized equivalent of analysis applied to chemical components:

"Analytical chemistry is a scientific discipline which develops and applies methods, instruments and strategies to obtain information on the composition and nature of matter in space and time" [1.1].

The information can be obtained by a step-by-step procedure on a macroscopic or microscopic scale (see Fig. 1.1).

The mesoscopic region, between the macroscopic and microscopic scales, can be described in the case of identification by clusters and polymers, in the case of localization as the nanometer region and in the case of characterization by the transition to the quantum regime. As regards quantification, the Avogadro number provides the connection between the macroscopic and microscopic scales.

The identification (first step of the scheme) defines the so called **analyte** by means of chemical, electrochemical, spectroscopical and physical properties. On the microscopic scale, identity can be defined by considering the electronic structure, the nuclear structure and the molecular structure (see Fig. 1.2).

For practical reasons, the definition of the analyte on the macroscopic scale is sometimes less strict and in these cases depends on the selectivity of the analytical procedure including the extraction (specification). All quantities and properties on the macroscopic scale can be represented by standards. In many cases these standards are divisible without changing the properties. Then they are called reference materials. The so-called reference materials have to be characterized by both homogeneity and stability. The life time of a reference material is limited not only by its stability, but also by scientific or

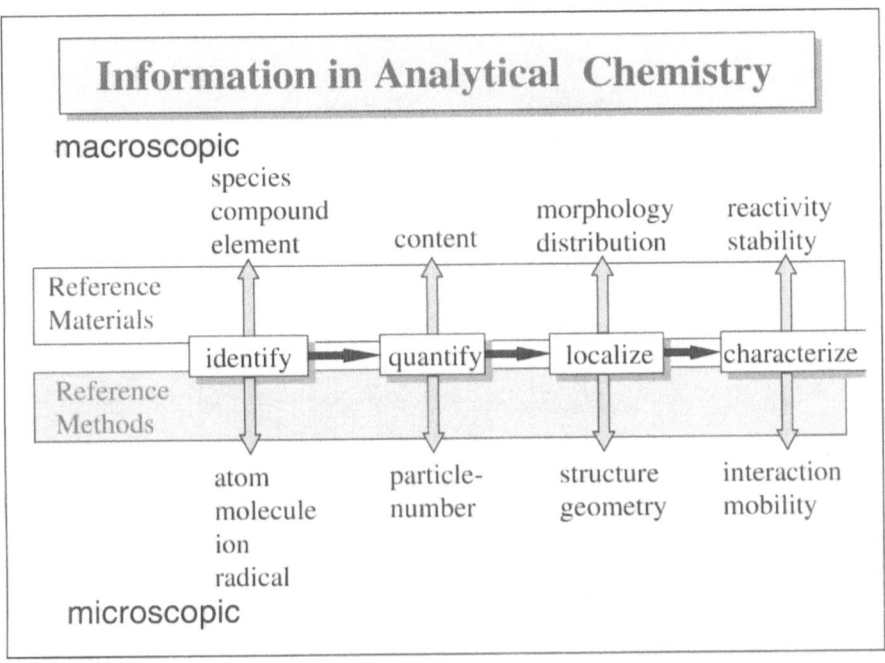

Fig. 1.1. The step-by-step acquisition of information in analytical chemistry

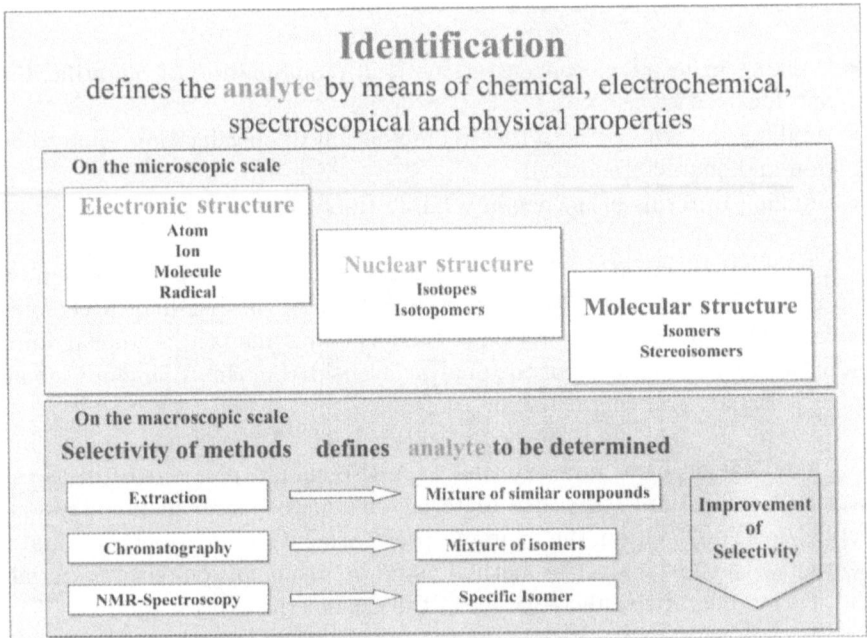

Fig. 1.2. Identification on microscopic and macroscopic scales

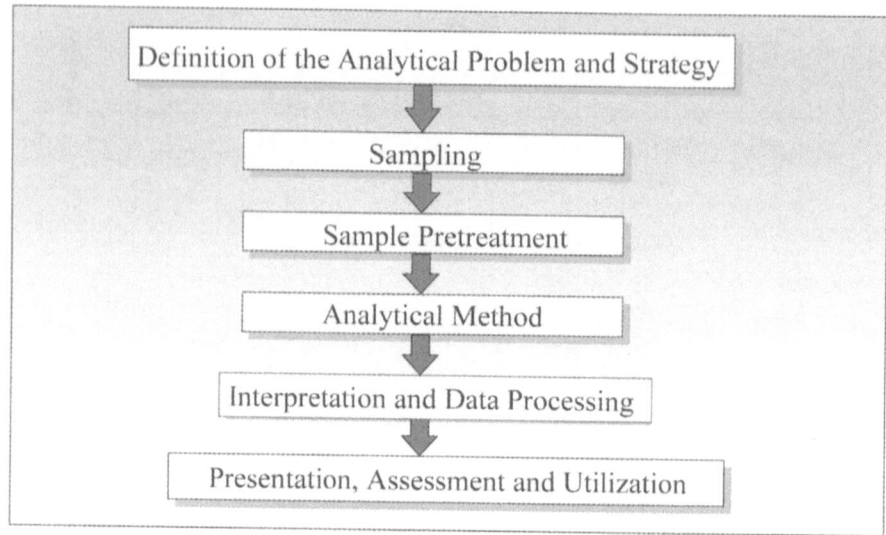

Fig. 1.3. The step-by-step procedure of chemical analysis

analytical progress. Reference materials are only valid in a reference system described by the analytical problems, the defined analytes and the measuring system.

Analytical chemical thinking is:

- thinking in terms of concentrations [1.2] (combination of quantification and localization);
- thinking in terms of activities (combination of specification, quantification and characterization);
- thinking in terms of uncertainty limits (metrological aspect).

Analytical chemical methodology follows an other dimension of step-by-step procedure (see Fig. 1.3). Each step influences the result of a chemical analysis. The correct consideration of chemical laws in all these steps requires the professional experience of the analyst. Considering this Wilhelm Ostwald has called analytical chemistry an art [1.3].

However an analysis is never an end in itself. An external need defines the analytical problem and external expertise should participate in assessment and utilization. Reference materials are mainly used in the step "analytical method". Often the "sample pretreatment" step can be evaluated together with the "analytical method" step by means of reference materials. The scheme must be slightly varied in the case of the process analysis.

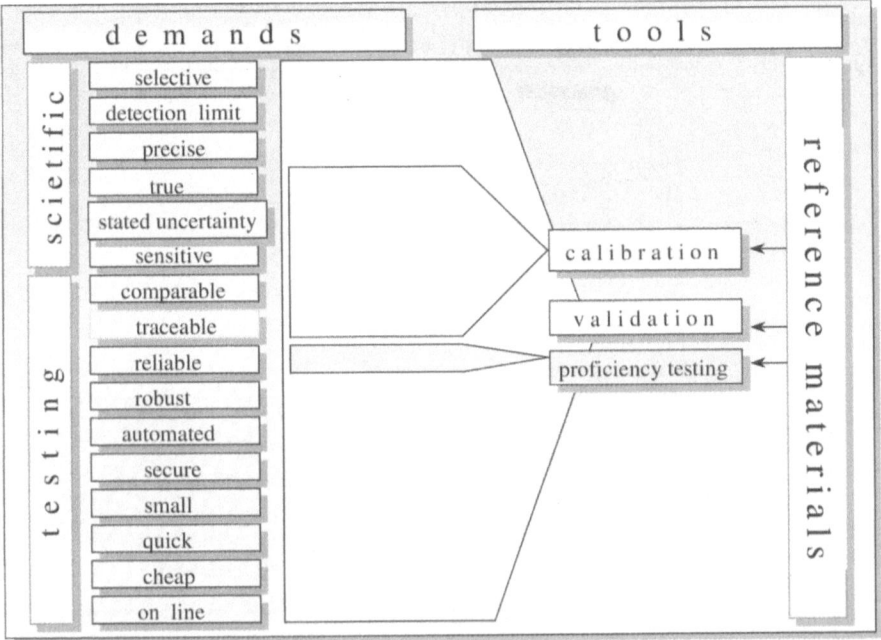

Fig. 1.4. Demands on analytical chemistry from the view of the scientific discipline and the testing field (customers)

1.2   Demands on Analytical Chemistry

Analytical chemistry with its testing results creates a basis for decisions in economics, politics, environmental protection, health service, trade, administration of justice and sports. The demands on analytical chemistry differ to some extent depending on whether they come from purely scientific sources or arise in the field of testing. In the latter case the customer defines the demand. The gathered demands have partly overlapping meaning and are not independent of each other (see Fig. 1.4).

Securing at the same time selectivity, precision and sensitivity [1.4] is unachievable for the same reasons as the simultaneous securing of reliability, rapidity and cheapness. There is always an optimum in the fulfilment of a certain number of demands that makes chemical analysis fit for purpose. Quality is the degree of fitness for purpose. Tools to assess fulfilment of demands are calibration, validation and proficiency testing. These are important elements in quality assurance (QA) and quality control (QC) [1.5] as well as being parts of the accreditation [1.6]. Reference materials are needed for all these procedures [1.7–11]. They establish the traceability of analytical results.

The range of reference materials spreads across a three-dimensional coordinate space: analytes, matrices and applications (see Fig.1.5). Systems of classification (following application fields) are given by reference materials

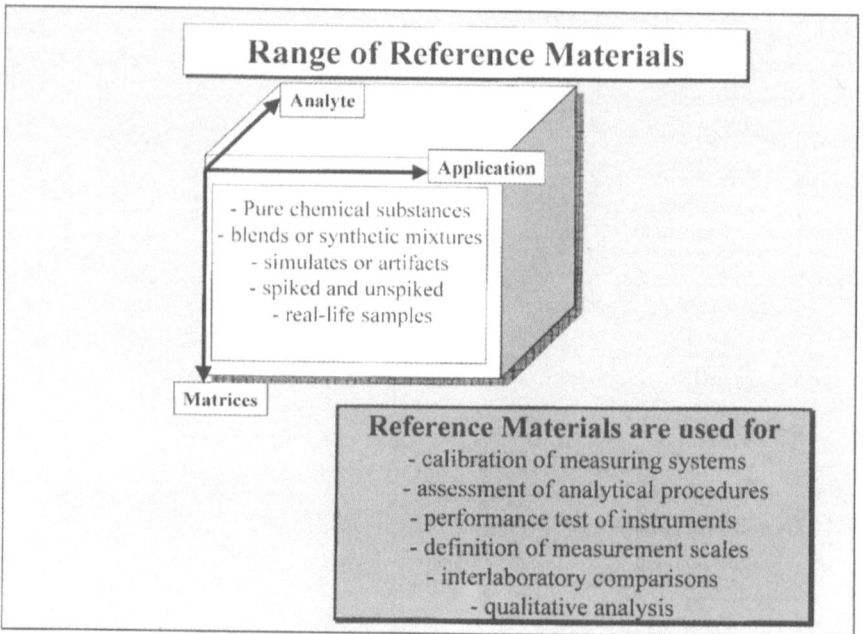

Fig. 1.5. Range of reference materials

catalogues of bodies such as the National Institute of Standards and Technology (NIST), the Laboratory of the Goverment Chemist (LGC), the Database for Certified Reference Materials (COMAR), the International Laboratory Accreditation Cooperation (ILAC) and the European Community Bureau of Reference (BCR).

1.3   Measures Designed to Build Confidence

The importance of reference materials for the reliability of analytical results demands a high degree of confidence in the quality of the reference materials. The certificate of a reference material and possibly an additional report reveal important information on quality aspects:

- property values and their uncertainty ranges;
- traceability information (conformity to a GUM-type uncertainty assessment [1.12]);
- measurement techniques, the use of primary methods and the procedure for data evaluation;
- instructions for the correct use of the reference material;
- stability, transportation and storage instructions;
- state of homogeneity (minimum sample size);
- declaration of conformity with standards and guides (e.g. ISO Guides 30–35 [1.13–18], BCR Guidelines [1.19]).

The availability of a certification report containing even more details of the certification procedure would be an additional aspect of confidence in the quality of the material.

A particularly important criterion will be the confidence in the organization which is responsible for the certification of the reference material. Sometimes the organization which is responsible for the supply of both the reference material and authorization of the data accompanying the reference material (certificates) is called the producer. Confidence in the producer is based on experience (proven by international cooperation, tradition and publications) and on competence (proven by interlaboratory key comparisons and expert reputation). A significant instrument to provide confidence could be the third party assessment of the competence of the reference material producer by accreditation, following ISO Guide 34 [1.17] or other specific requirement documents.

1.4   References

1.1. Federation of European Chemical Societies (FECS) Edinburgh (1993)

1.2. Davis R S, Brief Review of Quantities Describing Compositions of Mixtures, CCQM/98–9

1.3. Ostwald W (1920) Die wissenschaftlichen Grundlagen der Analytischen Chemie, Th. Steinkopff, Leipzig

1.4. Valcarcel M, Rios A (1997) Fresenius J. Anal. Chem. **357**, 202–205

1.5. EURACHEM/CITAC Guide CG 2 (1998) Quality Assurance for Research and Development and Non-Routine Analysis

1.6. ISO/IEC FDIS 17025 (2000) General Requirements for the Competence of Testing and Calibration Laboratories

1.7. Seward R W (Ed.) (1975) Standard Reference Materials and Meaningful Measurement, National Bureau of Standards (NBS), U.S. Dept. of Commerce, Washington, D.C.

1.8. Uriano G A (1979) The Use of Standard Reference Materials for Quality Assurance of Environmental Measurements, Inform. Transf. Inc, Silver Spring, MD, USA

1.9. Schmitt B F (Ed.) (1980) Production and Use of Reference Materials, Bundesanstalt für Materialprüfung (BAM), Berlin, Germany

1.10. Taylor I K (1993) Handbook for Standard Reference Materials Users, National Bureau of Standard (NBS), U.S. Dept. of Commerce, Washington, D.C.

1.11. Clement R E, Keith L H, Siu K W M (1997) Reference Materials for Environmental Analysis, Lewis Publ., Boca Raton, N.Y., London, Tokyo

1.12. ISO (TAG 4) WG 3 (1992) Guide to the Expression of Uncertainty in Measurement (GUM); EURACHEM ICITAC Guide (1995) Quantifying Uncertainty in Analytical Measurement

1.13. ISO Guide 30 (1992) Terms and Definitions used in Connection with Reference Materials

1.14. ISO Guide 31 (1981) Contents of Certificates of Reference Materials

1.15. ISO Guide 32 (1996) Calibration in Analytical Chemistry using Reference Materials

1.16. ISO Guide 33 (1989) Use of Certified Reference Materials

1.17. ISO Guide 34 (1996) Quality Systems Guideline for Production of Reference Materials

1.18. ISO Guide 35 (1989) Certification of Reference Materials – General and Statistical Principles

1.19. Doc. BCR /48/93 (1993) Guidelines for the Production and Certification of BCR Reference Materials

2 Classification of Reference Materials

Werner Hässelbarth

Reference materials can be classified according to various aspects. The classifications considered in this chapter, however, are far from the systematic categorizations as available for organic compounds, for example. Instead, the intention is to outline major aspects bearing on analytical practice.

- *Physical Character.* Reference materials can be gases, liquids and solids. Each state of matter has its specific features and problems concerning preparation and handling of reference materials. In addition, solid reference materials offer the possibility of considering bulk characteristics, localized characteristics and spatial distribution characteristics.
- *Supplied Property.* Reference materials for analytical chemistry are designed to realize one or several properties. These can be a pure chemical species, the composition of a mixture, or a physico-chemical property. In addition, as already mentioned, for solid materials there is, at least in principle, a smooth transition from bulk reference materials to reference objects.
- *Metrological Qualification.* Reference materials are the main building blocks of traceability chains in analytical chemistry. As such, their metrological characteristics, in particular the uncertainty of supplied properties and the position in a traceability hierarchy, are items of focal interest for quality assurance.
- *Preparation Method.* Reference materials can be prepared according to different procedures, principles or even "philosophies". Gaseous and liquid materials are most often prepared synthetically. Solid reference materials, however, are burdened with strong matrix effects. Therefore, in particular in environmental analysis, analyzed natural materials are preferred over synthetic materials.
- *Intended Use.* Reference materials are mainly used for calibration of analytical instruments, i.e. determination of the relationship between instrument response and analyte content, and validation of analytical methods, i.e. assessment of method performance, in particular accuracy. Identification of analytes is another important field of application. Recently "proficiency testing materials", designed for use in interlaboratory performance studies, have gained considerable attention.

These topics will be considered in more detail in the sections to follow. However, the intention is to present not a comprehensive, systematic treatment but rather an introduction to "horizontal issues" occurring in various application fields, as covered in later chapters.

First, however, definitions of the fundamental terms "reference material", "certified reference material" and related terms are given in Sect. 2.1. As a common feature, all these definitions are constructed according to the substitution principle. There the aim is to condense into a single half sentence, that could be substituted for the defined term, all major characteristics of the defined object. Additional information is given in notes. For the convenience of the reader, the definitions from ISO Guide 30 are reproduced including notes, while the notes are omitted for the additional definitions from GUM and VIM.

2.1 Definitions

The fundamental terms used in connection with reference materials are defined in ISO Guide 30 [2.1]. For the purpose of this presentation, a selection of definitions is reproduced here. In addition, three basic terms from metrology are reproduced from the Guide to the Expression of Uncertainty in Measurement (GUM) [2.2] and the International Vocabulary of Basic and General Terms in Metrology (VIM) [2.3].

Reference Material (RM). Material or substance one or more of whose property values are sufficiently homogeneous and well established to be used

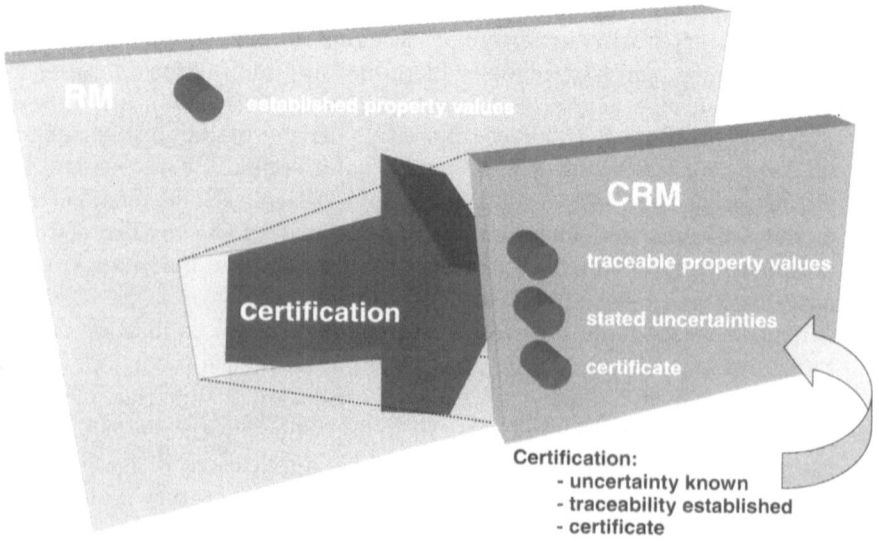

Fig. 2.1. Definition of RMs and CRMs

for the calibration of an apparatus, the assessment of a measurement method, or for assigning values to materials.

Note: A reference material may be in the form of a pure or mixed gas, liquid or solid. Examples are water for the calibration of viscosimeters, sapphire as a heat-capacity calibrant in calorimetry, and solutions used for calibration in chemical analysis.

Certified Reference Material (CRM). Reference material, accompanied by a certificate, one or more of whose property values are certified by a procedure which establishes its traceability to an accurate realization of the unit in which the property values are expressed, and for which each certified value is accompanied by an uncertainty statement at a stated level of confidence.

Note 1: Refers to the location of the definition of the term "reference material certificate".

Note 2: CRMs are generally prepared in batches for which the property values are determined within stated uncertainty limits by measurements on samples representative for the whole batch.

Note 3: The certified properties of reference materials are sometimes conveniently and reliably realized when the material is incorporated into a specially fabricated device, e.g. a substance of known triple point into a triple-point cell; a glass of known optical density into a transmission filter; spheres of uniform particle size mounted on a microscope slide. Such devices may also be considered as CRMs.

Note 4: All CRMs lie within the definition of measurement standards or etalons given in the International Vocabulary of Basic and General Terms in Metrology (VIM).

Note 5: Some RMs and CRMs have properties which, because they cannot be correlated with an established chemical structure or for other reasons, cannot be determined by exactly defined physical and chemical measurement methods. Such materials include certain biological materials such as vaccines to which an international unit has been assigned by the World Health Organization.

Certification of a Reference Material. Procedure that establishes the value(s) of one or more properties of a material or substance by a process ensuring traceability to an accurate realization of the units in which the property values are expressed, and that leads to the issuance of a certificate.

Reference Material Certificate. Document accompanying a certified reference material stating one or more property values and their uncertainties, and confirming that the necessary procedures have been carried out to ensure their validity and traceability.

Uncertainty of Measurement. Parameter, associated with the result of a measurement, that characterizes the dispersion of values that could reasonably be attributed to the measurand.
(Definition taken from the GUM [2.2] where it has three notes.)

Traceability. Property of the result of a measurement or the value of a standard whereby it can be related to stated references, usually national or international standards, through an unbroken chain of comparisons all having stated uncertainties.
(Definition taken from the VIM [2.3] where it has three notes.)

Measurement Standard (Etalon). Material measure, measuring instrument, reference material or measuring system intended to define, realize, conserve or reproduce a unit or one or more values of a quantity to serve as a reference.
(Definition taken from the VIM [2.3] where it has six examples and three notes.)

According to the definitions of ISO Guide 30, as given above and summarized in Fig. 2.1, CRMs are a subclass of RMs, i.e. a RM can be a CRM or a non-CRM. However, often the term RM is (mis)used to denote exclusively non-CRMs.

2.2   Physical Character

Reference materials can be gases, liquids or solids. Figure 2.2 shows some examples from the range of reference materials developed by BAM. Each state of matter has its specific features and problems concerning preparation and handling of reference materials. In addition, solid reference materials offer the possibility of considering bulk characteristics, localized characteristics and spatial distribution characteristics.

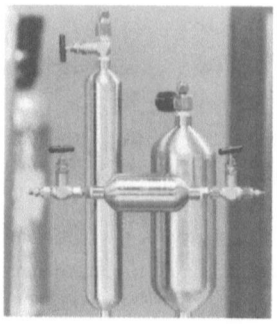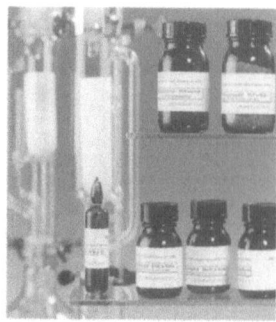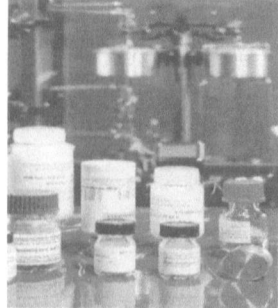

Fig. 2.2. Examples of gaseous, liquid and solid reference materials

Gaseous reference materials, most often called reference gases or calibration gases, are mainly applied in gas analysis, which includes analysis of gas mixture composition, analysis of trace impurities in pure gases, and measurement of physico-chemical gas properties such as the calorific value of gaseous fuels. When entering this field of analytical chemistry, one has to get used to the feature that gases are volatile and almost immaterial substances, which are handled exclusively in closed systems.

Reference gas mixtures are normally prepared synthetically, by mixing measured amounts of pure gases or of gas mixtures of known composition. For this purpose, two different preparation strategies are in use: *static methods* for preparing a stock of material in a container, e.g. the gravimetric method, where the mixture components are transferred consecutively into a pressure cylinder and the amount is determined by weighing; and *dynamic methods* for preparing transient samples for immediate use, e.g. dynamic volumetric methods, where the mixture components are conducted, at controlled volume flow rates and constant pressure, into a mixing tube from where the mixture is transferred to the point of use.

Using these techniques, gas mixtures can be prepared with high accuracy. However, great care is required in handling and transfer, to ensure that at the time and point of use the mixture composition conforms with the preparation recipe. Mechanisms for mixture degradation are numerous, including diffusion of air into the gas stream, loss of gas through leakage, selective adsorption of mixture components by vessel and transfer line materials and condensation of low-vapour pressure components.

Liquid reference materials are most often aqueous solutions, containing defined amounts of single or several specified analytes, e.g. heavy metal ions. Their typical use is calibration of analytical instruments, e.g. atomic adsorption spectrometers. Calibration solutions are, however, by no means confined to elemental analysis. Among many others, solutions of trace organic compounds in organic solvents are used in environmental analysis, solutions of serum materials in clinical analysis, and solutions of inorganic salts in measurement of electrolytic conductivity.

As a rule, calibration solutions are prepared synthetically, by dissolving measured amounts of pure substances in measured amounts of solvent. The main sources of potential error to be mastered are impurities of the parent substances and the solvent.

Similar to gas mixtures, calibration solutions can be prepared with high accuracy, while problems may arise in storage and handling. Among others, calibration solutions may suffer from contamination by air components (e.g. carbon dioxide will affect any standard base solution), contamination by leaching from vessel materials, evaporation of solvent (e.g. at ambient temperature aqueous solutions in plastic bottles may exhibit significant upgrading due to diffusion of water vapour; see e.g. [2.4]), and degradation of unstable analytes.

Solid reference materials come in an enormous variety of forms, ranging from metal disks to milk powder. Each form of solid RMs – massive samples, chips, granules, powders etc. – gives rise to specific problems in the quest for generating representative and homogeneous samples (see Chap. 4). For example, massive samples of metal alloys are prone to inhomogeneities caused by segregation during crystallisation of the melt, while powders often behave like highly viscous fluids and therefore are difficult to mix. In addition, solid reference materials are not only designed to supply bulk properties, but also localized properties such as the composition of surface layers, or spatial distribution properties such as the pore size distribution in porous materials. The application range of solid reference materials is equally wide, going from analysis of metals to food analysis and including measurement of various physico-chemical properties. Though localized properties are of growing importance, the present introduction is restricted to characteristic features of solid reference materials for bulk chemical composition.

Unlike gas mixtures and solutions, where synthetic mixtures dominate, solid reference materials are almost without exception "natural materials" – i.e. refined samples of typical industrial, environmental or biological materials – whose composition is determined by analysis (for a noticeable exception see [2.5]). Most often, this analysis is carried out as an interlaboratory study, by several expert laboratories, using various independent analytical methods.

Concerning preparation and use, solid reference materials also give rise to problems not encountered with gaseous and liquid materials, which are due to inhomogeneity. That is, the analyte content, e.g. the content of chromium in steel or the content of PCBs in a river sediment, are not necessarily equidistributed throughout the material. Therefore, samples may differ in bulk analyte content, and this inhomogeneity will increase with decreasing sample size. Preparation of solid reference materials therefore requires great care to ensure that the material is sufficiently homogeneous, and major efforts in the subsequent analysis are devoted to homogeneity testing. Potential inhomogeneity also has to be taken into account in the use of solid reference materials, e.g. concerning sample intake for analysis. Reference material certificates of potentially inhomogeneous materials will normally specify the minimum sample intake, in relation to the specified uncertainty.

At first sight, stability may appear to be a minor issue in comparison with reference gas mixtures and calibration solutions. However, solid materials may also suffer from degradation, e.g. through chemical reactions between analytes and matrix constituents, decomposition of unstable analytes and microbiological activities. This applies in particular to materials which are certified for the content of organic compounds and microbiologically active materials such as soil samples.

2.3 Supplied Property

Reference materials for analytical chemistry are designed to realize one or
several properties. These can be a pure chemical species, the composition of
a mixture or a physico-chemical property. In addition, as already mentioned,
for solid materials there is, at least in principle, a smooth transition from
bulk reference materials to reference objects.

From a systematic point of view, before considering reference materials
for chemical composition, realizations of the analytes under consideration are
needed first. This also makes sense from a practical view, since often enough
a compound determined with high accuracy has later turned out to be the
wrong one. Quantification without proper identification is a waste of effort
and, in addition, can have disastrous consequences. Pure substances are also
needed for preparing calibration solutions and other synthetic mixtures.

Pure substances are, of course, a fiction. What is needed in reality are
substances where the total content of impurities is very low and, equally important, safely known. This condition, however, is not easy to fulfill in practice. This is due to the fact that direct determination of the main component,
say the mass fraction of copper in a purefied copper material on a level of
99.999%, will usually lack the required accuracy, which would be 10^{-6}. The
only way out is therefore to identify all potential impurities, determine these
and subtract from 100%. This approach, however, has the drawback that, at
least in principle, significant impurities may be overlooked, as has happened
with gases in metals.

The most important type of supplied property is mixture composition,
that is, the content of one or several specified analytes in a mixture. Evidently
composition reference materials play the role of measurement standards for
composition measurements, i.e. for quantitative chemical analysis. Complete
mixture composition is rarely required, a prominent exception being natural
gas, where the complete composition is used for determining the calorific
value. As a rule, only the content of a single or a few specified analytes is
determined, together with a qualitative or semi-quantitative specification of
the matrix. Composition reference materials are designed correspondingly, as
realizations of reference compositions relevant for analytical practice.

Reference materials are, however, also used to realize various reference
values of physico-chemical properties. Prominent examples are: heat of combustion (benzoic acid), fixed points of the International Temperature Scale
(zinc, tin and other metals), and fixed points on the IUPAC pH scale (buffer
solutions). In addition, solid reference materials can be used to realize localized properties such as the composition or other properties of surface layers,
or spatial distribution properties such as the pore size distribution in porous
materials. Items such as hardness blocks would rather be called reference objects than reference materials. Anyway, there is a wide range between bulk
materials from which aliquots are sampled for measurement and artifacts
which are used as an entity.

Much finer subdivisions concerning types of supplied properties have been developed in reference material catalogues of bodies such as the National Institute of Science and Technology (NIST) and the European Community Bureau of Reference (BCR) and in the COMAR database (see Chap. 8). In a recent ILAC document [2.6] these classifications have been amalgamated into a detailed classification system, starting with five principal categories as follows:

Category A: Chemical Composition. Reference materials, being either pure chemical compounds or representative sample matrices, either natural or with added analytes (e.g. animal fats spiked with pesticides for residue analysis), characterized for one or more chemical or physico-chemical property values.

Category B: Biological and Clinical Properties. Reference materials similar to Category A, but characterized for one or more biochemical or clinical property values, e.g. enzyme activity.

Category C: Physical Properties. Reference materials characterized for one or more physical property values, e.g. melting point, viscosity, density.

Category D: Engineering Properties. Reference materials characterized for one or more engineering property values, e.g. hardness, tensile strength, surface characteristics.

Category E: Miscellaneous Properties.

These principal categories are subdivided into sub-categories, employing up to three additional levels. As an example, an aluminium alloy characterized for the content of minor components such as Mn, Si, Cu, Ni and Cr would be listed under chemical composition – metals – nonferrous metals – aluminium alloys.

2.4   Metrological Qualification

Reference materials are the main building blocks of traceability chains in analytical chemistry. As such, their metrological characteristics, in particular the uncertainty of supplied properties and the position in a traceability hierarchy are items of focal interest for quality assurance.

Concerning the first qualifier, the uncertainty of supplied properties, there are two main requirements: the uncertainty has to be known and fit for purpose. Reference materials may therefore be ranked according to the uncertainty level (the lower the better), and according to the level of reliability and authentication of the uncertainty statement.

In physical metrology, measurement standards are commonly classified according to levels in a hierarchy as follows. Primary standards of measurement constitute the top level with respect to both qualifiers. Then secondary standards are derived or verified by direct comparison with one or several primary standards, and so on. This hierarchical system is used to establish the accuracy of measuring systems. These are calibrated using working standards, which in turn have been calibrated against reference standards, and so forth. Ideally, all these traceability chains terminate at the level of primary standards. By this procedure, bias of the measuring system is corrected, if significant, and the measurement uncertainty is traced back to the uncertainty of the primary standard(s) and the uncertainty of the comparison measurements involved.

In analytical chemistry an analogous system has not been established so far, with the exception of specific fields in gas analysis (e.g. automotive exhaust measurements). This has to do with the fact that comparison of analytical standards usually implies sub-sampling and sample pre-treatment, while physical standards, e.g. two gauge blocks, can often be compared directly. Therefore in analytical chemistry the uncertainty associated with traceability links is often much higher, with the consequence that long traceability chains are ineffective. At present, the only widely accepted classification of reference materials bearing upon metrological aspects is between "certified reference materials", as defined in Sect. 2.1, and "in-house reference materials", which is the common name for reference materials prepared by users for their own purposes. These two classes clearly differ with respect to the level of authentication, but not necessarily with respect to the level of uncertainty.

Recently, however, considerable efforts have been made to establish the concept of "primary reference materials" as an analogue of primary standards of measurement. The starting point for this initiative was the foundation of the Comité Consultatif pour la Quantité de Matière (CCQM) in 1993, as a new committee of the Meter Convention, related to the SI unit "mole", with the task of developing concepts, methods and structures for top-level metrology in analytical chemistry. As one of its first actions, the CCQM adopted a general definition of a "primary method of measurement", intended to serve as a more accurate definition of what has often been called an absolute method or a definitive method [2.7]. Primary reference materials were then defined as reference materials whose values were determined using a primary method.

The formal definitions adopted by the CCQM are as follows [2.8]:

- "A primary method of measurement is a method having the highest metrological qualities, whose operation can be completely described and understood, for which a complete uncertainty statement can be written down in terms of SI units, and whose results are, therefore, accepted without reference to a standard of the quantity being measured."

- "A primary reference material is one having the highest metrological qualities, and whose value is determined by means of a primary method."

At present, however, for certified reference materials the (certified) value is normally determined in an interlaboratory study, as a consensus value derived from the results obtained by different expert laboratories using several independent methods. There is no reason why this procedure should not qualify, in principle, for the generation of primary-level reference materials. Crucial conditions, however, would be that only validated methods with specified uncertainty are admitted and that a consensus value is only assigned if the individual results agree within their specified uncertainty limits. Recently this view has also been largely accepted by the CCQM. In addition, the concept of a "primary method of measurement" has to be complemented by that of a "primary method of preparation", applicable to the preparation of mixtures or solutions from pure substances.

Concerning terminology, the term "primary method" has largely been accepted by the analytical community as a refinement of the term "definitive method". The term "primary reference material", however, is still under discussion, for example often being restricted to pure substances. For this reason, the term "primary reference material" is not used in this book.

In physical measurements it is common practice to calibrate or verify working standards by comparison with corresponding reference standards. Analogously certified reference materials could (and should!) be used as reference standards for establishing traceability of in-house reference materials used as working standards. For this purpose, protocols have to be developed, where, as an important item, the propagation of uncertainties from certified reference materials to in-house reference materials has to be given particular attention (for a recent example see [2.9]).

2.5   Preparation Method

Reference materials can be prepared according to different procedures, principles or even "philosophies". Gaseous and liquid materials are most often prepared synthetically. Solid reference materials, however, are burdened with strong matrix effects. Therefore, particulary in environmental analysis, analyzed natural materials are preferred over synthetic materials.

Preparation of a pure-substance reference material involves two main steps. First an appropriate raw material is purified, i.e. impurities are removed. As a rule, high-level purification requires highly specialized techniques, tailored for the material, the impurities and the impurity level concerned. Second, the content of residual impurities is specified. This is done for two different purposes. As already mentioned before, direct determination of purity, i.e. of the content of the main component, say the mass fraction of copper in a purefied copper material, will usually lack the required accuracy. Therefore purity is almost always determined indirectly, i.e. identified impurities are determined and subtracted from 100%. In addition, pure substances with quantified impurites may be used as reference materials for trace analysis.

Gaseous reference materials are, almost without exception, synthetic mixtures, prepared in batches for stock, or prepared continuously for immediate use. In principle, analyzed samples of "naturally occurring" gas mixtures could serve the same purpose. The main reasons for preferring synthetic mixtures are as follows:

- natural mixtures most often contain components not wanted in the mixture, which may adversely affect the stability of the mixture or interfere with the determination of specified analytes in intended applications;
- for natural mixtures it will often not be practical to establish the complete composition, i.e. the content of all components down to the trace level;
- stationary mixtures (i.e. mixtures in pressurized bottles) can be prepared at much higher accuracy (0.1% relative and better), using the gravimetric method, than the accuracy available by analysis;
- mixtures with analytes at extremely low levels or reactive analytes cannot be kept sufficiently stable in containers and therefore have to be prepared "on-line".

As a rule, synthetic solutions are also preferred over natural solutions, for largely the same reasons as explained above: no unwanted components, complete command of composition, and higher accuracy.

For solid reference materials, however, the situation is completely reversed. Almost without exception, reference materials are analysed samples of "natural" materials. This preference is mainly due to the limited selectivity of analytical methods for solid materials, in particular concerning the performance of sample preparation procedures. As a rule, analytical methods for solid materials are matrix-sensitive. For method validation, therefore, reference materials are required whose matrix does not differ significantly from the matrix of the samples to be analysed. Most often such materials – e.g. soil samples, food samples, and biological materials – cannot be prepared artificially. But even for manufactured reference materials such as metal alloys, the composition is normally determined by analysis instead of being derived from mixture formulation. Only recently, "re-constitution analysis", i.e. controlled preparation and application of solid samples from pure materials for quality assurance of multi-element methods (e.g. XRF), has been described in a monograph [2.5]. Concerning preparation of solid reference materials for bulk properties, one of the main challenges is to generate representative and homogeneous samples. As already mentioned before, solid reference materials come in an enormous variety, where each form – massive samples, chips, granules, powders, etc. – gives rise to specific problems. For example, massive samples of metal alloys are prone to inhomogeneities caused by segregation during crystallisation of the melt, while powders often behave like highly viscous fluids and are therefore difficult to mix. These topics occur throughout the sections on application fields of reference materials, see e.g. Chap. 4.

Indisputably, reference materials are the main tool for quality assurance in analytical chemistry. However, provision of tailor-made reference materials for every analytical task will clearly be impossible. Instead, more attention and effort should be given to the development of reference methods. Apparently, the "work-horse" analytical methods will continue to be plagued with matrix sensitivity. But then, reference methods of high selectivity would evidently be much better suited than reference materials for assessing the accuracy of routine methods, since the problem of matrix mismatch does not arise. In this approach, a reference method of known accuracy is applied in parallel with the routine method to real-life samples, and the results are compared.

2.6 Intended Use

Reference materials are mainly used for calibration of analytical instruments, i.e. determination of the relationship between instrument response and analyte content, and validation of analytical methods, i.e. assessment of method performance, in particular accuracy. Recently "proficiency testing materials", designed for use in interlaboratory performance studies, have gained considerable attention.

The basic function of reference materials in analytical chemistry is to supply reference values of chemical composition or physico-chemical properties. Here it is important to notice that for proper use of these reference values their uncertainty has to be known. In the paragraphs to follow the main lines of use are outlined. For further reading, ISO Guide 33 [2.10] is recommended as the most widely accepted generic guidance document on uses of reference materials.

2.6.1 Calibration of Measuring Systems

In composition analysis by instrumental methods the target quantity, i.e. the content of a specified analyte in a specified matrix, is most often not measured directly. Instead, an instrumental response is measured, which has to be converted into an analyte content. For determining the relationship between instrumental response and analyte content, the response is measured on calibration samples of known analyte content, covering the measuring range. From the comparison between the measured responses and the reference values of the analyte content the parameters of the response curve (e.g. slope and intercept of a straight line) are derived, including uncertainties in these parameters. By means of these data the analyte content of an unknown sample can be predicted from its measured response, and from the uncertainty of the measured response and the uncertainties of the response curve parameters the uncertainty of the predicted analyte content can be calculated. ISO

11095 [2.11] gives a general description of the design of calibration experiments using reference materials and the evaluation of calibration data for the common case where the calibration curve is a straight line.

The uncertainty budget derived from calibration is, however, incomplete if the analysed samples significantly differ in matrix from the calibration samples, as will most often be the case. Then matrix-matched samples have to be used, in addition, to determine, and eventually correct, bias due to matrix mismatch in calibration. In principle, matrix-matched reference materials could already be used for calibration, but in practice this is only done in special fields, e.g. in gas analysis. Usually calibration is performed using calibration solutions prepared from pure substances, and matrix-matched reference materials are used to investigate potential bias due to matrix influences. This practice has even led to a general belief that matrix reference materials cannot be used for calibration, but only for quality control.

2.6.2 Assessment of Measuring Methods

As part of a method validation study, potential bias is determined, and corrected if significant, using reference materials whose matrix closely matches that of the samples to be analyzed, most often a single validation sample only, due to the lack of suitable matrix reference materials. For determining potential bias, the analytical method is applied to the validation samples, and the results are compared with the reference values attributed to the reference materials used. Alternatively, the analytical method under investigation and a reference method (i.e. a method whose uncertainty is known and sufficiently small) are applied in parallel to appropriate samples, and the results of the candidate method are compared with those of the reference method.

This comparison is performed and evaluated in a series of three steps as follows:

- measurement results and corresponding reference values are compared, taking into account the relevant uncertainties;
- if significant bias was found, a bias correction is applied, preferably by adjustment of the procedure or instrumentation, or by calculation using a correction factor or an additive correction term, for example;
- the uncertainty of the results of future measurements, including bias correction, is calculated by combination of two contributions: the uncertainty of the uncorrected measurement result and the uncertainty of the correction.

By this procedure, the accuracy of the analytical method is traced back to the reference values, including their uncertainties, attributed to the validation samples. These data, therefore, have to be well established by themselves (e.g. certified values) or traceable to acknowledged reference standards.

2.6.3   Definition of Measurement Scales

Measurement scales have been harmonized internationally to a great extent by adoption of the International System of Units (SI). Nevertheless, there are measurements whose results cannot be expressed in SI units, e.g. hardness. For these measurements, conventional measurement scales have to be used. These scales are based on values assigned to reference materials (or reference objects), in conjunction with a specified measurement method or apparatus. It is evident that the reference materials only define a selected number of points on the scale, the so-called "fixed points". The definition of a scale therefore also requires specification of an interpolation procedure.

Measurement scales based on reference materials are also used in case of SI quantities which are difficult to measure directly. As a prominent example, the International Temperature Scale (ITS-90) is based on a series of accurately reproducible thermodynamic equilibrium states of pure substances, among others the triple points of several pure gases and the melting points of a series of pure metals. The scale defined by these fixed points and the agreed interpolation procedure, in conjunction with a system of agreed measurement methods and apparatus, serves as a practical approximation to direct measurement of thermodynamic temperature.

Additional explanations and examples are given in ISO Guide 33 [2.10].

2.6.4   Identification of Analytes

As already mentioned in Sect. 2.3, a compound determined with high accuracy has often later turned out to be the wrong one. Proper identification of the analyte causing an observed response is therefore of paramount importance. Frequently, e.g. in toxicological analysis, identification is the main analytical task, and accurate quantification is less important (see e.g. Chap. 6). For this purpose, reliably characterized pure substances are required as reference materials to identify analytes through patterns of instrumental response.

A difficult task arising in this connection is the estimation of identification uncertainty, i.e. estimation of the probability of an erroneous identification. Typically, such estimation would have to combine different types of information, e.g. on likely candidate species, on response characteristics, and on structure-property correlation. To date, standard statistical tools for estimating identification uncertainty are still lacking. A recent account of the state-of-the-art and future perspectives in this field are given in [2.12].

2.6.5   Proficiency Testing

As noted in [2.13], a major trend producing demand for reference materials of a special type has developed over the past decade. This trend is towards increased use of proficiency testing in connection with laboratory acreditation. Basically, in proficiency testing, participants analyze identical blind samples, whose analyte content (or other property value) is however known to the organizer, and performance is assessed by comparison of participants results with the target value, using an appropriate performance measure such as the z-score (see [2.14]). For this purpose, the target value should be well established, e.g. from an independent reference analysis. However, the grand mean of all participant results has often been, and still is, used as a substitute for an independently established reference value. This precedure is by now clearly recognized as insufficient. As a consequence, PT samples with well established target values are increasingly demanded by organizers of PT schemes. A prominent example of an "SI-traceable" proficiency testing scheme is the International Measurement Evaluation Programme (IMEP) of the IRMM, Geel.

Evidently requirements on PT materials are basically the same as for reference materials: their property values have to be sufficiently well established, homogeneous and stable to be used for the intended purpose. However, larger uncertainty may be acceptable than for an analogous certified reference material, and proof of long-term stability is not required. Hence PT materials could either be prepared under a separate production line, or in conjunction with a CRM production line, where part of the material is released after preliminary stability testing, with values assigned in a preparatory interlaboratory study.

2.7 References

2.1. ISO Guide 30 (1992) Terms and definitions used in connection with reference materials

2.2. ISO (1993, 1995) Guide to the Expression of Uncertainty in Measurement

2.3. ISO (1993) International Vocabulary of Basic and General Terms in Metrology

2.4. Wampfler B, Rösslein M (1998), Accred. Qual. Assurance, Vol. **3**, 468–470

2.5. Staats G, Noack S (1996) Qualitätssicherung in der Analytik – Die Rekonstitution Verlag Stahleisen GmbH, Düsseldorf

2.6. ILAC (2000) Guidelines for the Requirements for the Competence of Reference Materials Producers

2.7. Cali J P, Reed W P (1976) in: Accuracy in Trace Analysis, NBS Special Publication 422, 41–63

2.8. Quinn T J (1997), Metrologia, **34**, 61–65

2.9. ISO/DIS 6143 (1998) Gas analysis - Determination of Composition and Checking of Calibration Gas Mixtures - Comparison Methods (Draft International Standard 1998)

2.10. ISO Guide 33 (1989) Uses of certified reference materials

2.11. ISO 11095 (1996) Linear calibration using reference materials

2.12. Ellison S L R, Gregory S, Hardcastle W A (1998), Analyst, **123**, 1155–1161

2.13. Rasberry S D (1998), Fresenius J Anal Chem, **360**, 277–281

2.14. ISO/IEC Guide 43-1 (1997) Proficiency Testing by Interlaboratory Comparison - Part 1: Development and Operation of Proficiency Testing Schemes

3 Certification of Reference Materials

Tomas Tamberg

3.1 Procedures and Strategies

3.1.1 The Certification Process and Its Aims

The definitions for the terms "certified reference material", "certification of a reference material" and "reference material certificate" are listed in Sect. 2.1.

In contrast to these more technical definitions, ISO Guide 34 [3.1] offers a definition for the reference material producer or certifier that refers back to other guides: *Technically competent body (organization or firm, public or private) that produces reference materials in accordance with ISO Guides 31 and 35* [3.2, 3.3].

It should be noted, that there is a sometimes confusing alternative use of the term "certification" in the sense of a confirmation of claimed qualities or already existing data by a third party. This purely confirmative procedure is called product certification [3.4]. In contrast to this the "certification of a reference material" implies the production, evaluation and assessment of original analytical data by the same party.

From the above-mentioned definitions the qualities that distinguish a CRM from other types of reference materials are:

- the availability of an uncertainty statement at a stated level of confidence with each certified property value,
- the existence of a certificate providing
  - all the information that is necessary for the correct use of the material,
  - some of the information that establishes confidence in the reliability of the certified values.

Furthermore, and as a kind of background to the requirements mentioned in the above definition, the certification process as it is understood today is based on a self-critical attempt to aim for certified values as near as possible to the true values and a transparent evaluation of the uncertainty ranges according to internationally accepted rules.

If the uncertainty evaluation is done in the proper way, its completion includes the proof that the certified property values are traceable to an accurate realization of the unit in which the property values are expressed.

Accuracy/Uncertainty: What Does It Mean Today?

The term "accuracy" which, when applied to a set of test results, involves a combination of random components (reproducibility) and a common systematic component (trueness), is defined in ISO 3534-1 [3.5] as *"the closeness of agreement between a test result and the accepted reference value"*. As a result of the worldwide acceptance of the concepts defined in the "Guide to the Expression of Uncertainty in Measurement(GUM)" [3.6], the positive term "accuracy" has been displaced in general use by the negative, but clearer term "uncertainty" which is defined as a *"parameter, associated with the result of a measurement, that characterizes the dispersion of the values that could reasonably be attributed to the measurand"*.

It is on the word "reasonably" that the procedures described in GUM and the efforts during a state-of-the-art certification procedure focus. There is no sense in certifying property values with a very small uncertainty range when there is a risk that the true value is outside this range with more than the stated probability. On the other hand there is no sense in certifying property values with artificially enlarged uncertainty ranges simply to be on the safe side. If a 95% confidence level is stated for an uncertainty range that in fact represents a 99.9% confidence level, the users of the CRM introduce an unjustified uncertainty component in their own uncertainty calculations.

Comparability

One of the major intentions in the use of reference materials is the improvement of the comparability of measurement results from different laboratories. Reference to a common standard of chemical composition or to standards traceable back to a common standard of higher order is a powerful tool to achieve comparability in a framework of international rules or between trade partners.

The term "comparability" has a floating definition [3.7] ranging between:

"The ability of measurement results to be compared, in order to determine whether they are equal or different (larger, smaller). This is achieved by expressing them on the same, preferably internationally agreed, measurement scale."

and:

"The property of measurement results, obtained on subsamples of the same material, to agree within their uncertainties, when the results are measured on the same measurement scale (i.e. expressed in the same unit)."

In a qualitative way, the latter definition is the one that most people have in mind when aspects of mutual recognition of analytical measurement results are considered.

Traceability

This term is defined as follows [3.8]:

"Traceability is the property of the result of a measurement or the value of a standard whereby it can be related to stated references, usually national or international standards, through an unbroken chain of comparisons all having stated uncertainties."

The definition of a CRM in Chap. 2 mentions the traceability of the certified values as an essential requirement for the status of such a material. However, traceability is not just a value in its own right; rather it is an essential precondition for establishing comparability of measurement results.

In the case of certified reference materials the certifier normally does not refer to other reference materials, but either directly to the SI units [kg] or [mol] or to written standards describing a measurement method. The direct reference to the international standard units [kg] or [mol] is achieved by introducing weighed quantities of sufficiently pure elements or compounds as calibrants into the analytical process. Calibration, therefore, is an important part of the certification procedure, but by no means the only important one.

The question to what extent weighed quantities of these pure elements or compounds can be considered as sufficiently "accurate realizations of the units in which the property values are expressed" has caused extended discussions between metrologists and analytical chemists.

For users, a certified reference material is the most important tool for achieving traceability of their own measurement results. They can even extend the traceability chain of comparisons downward by referring to the CRM in the process of defining secondary reference materials and working standards.

The Reference Material Certificate

One of the essential preconditions for the acceptance of a certified reference material is the availability of a certificate delivered with each unit of the CRM. The certificate must contain three types of information:

- description of the material;
- all information necessary for correct use;
- confidence-building information.

The following are the headings under which ISO Guide 31 requests information in the certificate to be supplied (if applicable):

- name and address of the certifying body,
- title of the document,
- name of the material,
- reference material code and batch number,
- description of the material,

- intended use,
- instructions for the correct use of the reference material,
- safety aspects,
- level of homogeneity,
- certified values and their uncertainty intervals,
- traceability,
- values obtained by individual laboratories or methods,
- uncertified values,
- date of certification,
- stability information,
- further information,
- legal aspects,
- signatures or names of certifying officers.

The Certification Report

The certification report is a more comprehensive document than the certificate, and it is usually not delivered with each unit of the CRM. (Often it is not even produced as a well structured document, but exists only as a collection of the papers on which all measurement results and decisions based on them have been documented.)

A good certification report should be structured in such a way that the individual steps of the certification process become apparent and that the decisions leading to the certified values and their uncertainty ranges become transparent.

If the certification report is not delivered with each unit of the RM, it should be made available on request to the RM user. If the report only exists as the above-mentioned collection of papers, the certifier should at least be able to use this documentation to answer quality related questions raised by users of the CRM.

Tasks and Responsibilities of the Certifier

The tasks of the certifier do not necessarily include running certification measurements. Certifiers may leave this task to a group of qualified laboratories and restrict their own activities to selecting the laboratories and/or methods, organizing and checking their work and evaluating their results. However, the certifier can in addition also act as one of the participating laboratories or provide all the participating laboratories.

Most of the following tasks can either be fulfilled directly by the certifier or delegated to others under his or her control:

- analysis of the needs for a new material and the number of possible users in comparison to production costs,

- planning of the certification procedure (including selection of the laboratories and methods involved[1]),
- procurement of the basic material,
- if necessary, processing of the basic material (e.g. drying, homogenization, purification, etc.),
- dividing the whole batch of material into units (e.g. bottling),
- testing for homogeneity and stability,
- organizing the certification measurements,
- evaluation of measurement results (must be done by the certifier; for details see Sect. 3.3),
- inducing necessary repetition measurements,
- writing the certificate and the certification report.

After certification, the certifier has the responsibility to inform all known users of the CRM about errors in the certification process that may cause withdrawal of the material or corrections to certified values and their uncertainty ranges. Furthermore, the certifier has to deal with the replacement of withdrawn units of the material.

3.1.2 Certification Strategies

Depending on accuracy targets, the difficulty involved in certification measurements and availability of analytical methods, there are different ways of structuring the system of certification measurements. However, the efficiency of the various approaches always depends on the quality of homogeneity and stability studies that have to be run prior to the certification measurements.

The Interlaboratory Cooperation Approach

This is a widely used concept, where the certifier has to find a group of laboratories with specific experience in the field of materials similar to that being analysed. The certifier will also aim for the use of as many independent analytical methods as possible by the group members. It is assumed that possible undetected systematic errors, leading to positive and negative biases, can in this way be levelled. The concept has the advantage that, in a field of 10 to 18 participants, for example, outliers can be identified and, after a technical discussion, be removed. The final result is mainly derived by treating the mean values from the single laboratories as monolithic values of equal merit and calculating a final mean value and a standard deviation from them.

Difficulties in the calculation of the uncertainties of the finally certified values have caused some metrologists to describe this approach as one that

---

[1] For the selection of laboratories and methods a qualification experiment with a similar material of known composition may be necessary.

produces non-traceable "consensus values". Therefore, intense work is currently underway in order to improve the statistical tools to be applied in this case [3.9]. On the other hand, comparison by the German Federal Institute for Materials Research and Testing (BAM) of the certification results of a series of interlaboratory cooperation projects with primary method results (IDMS) in the field of metal alloy CRMs has in no case shown essential contradictions.

The Elite Group of Methods Approach

Where a variety of analytical methods is available in one institution or in a small group of institutions, the certification can be done on the basis of the results from only two to four laboratories/methods provided that

- analytical methods are carefully validated,
- validation covers a range of sample composition similar to that of the candidate CRM,
- staff members applying the methods are well trained in doing so,
- methods are selected in such a way that the number of common sources of uncertainty is minimized,
- results are in good agreement.

This concept relies on long term experience of the analysts, careful conservation and documentation of the experience and well known status for the available instrumentation.

The Primary Method Approach (Primary Reference Materials)

In cases where the uncertainties in certified values have to be minimized, a single, so-called primary analytical method, can be used for certification. If this is done in only one laboratory, the additional application of another non-primary analytical method is recommended as a blunder check.

Unfortunately, the definition of the term "primary method" and the identification of methods which fulfill the requirements of that definition are far from being completed.

The current definition as issued by CCQM reads as follows:

"A primary method of measurement is a method having the highest metrological qualities, whose operations can be completely described and understood, for which a complete uncertainty statement can be written down in terms of SI units and whose results are, therefore, accepted without reference to a standard of the quantity being measured."

Explanatory notes:

1. *A primary method also requires an equation that connects what is measured with what is intended to be measured without including any significant empirical correction factors.*

2. *Measurements of amount of substance, to be considered primary, must be made using a method which is specific for a defined substance and for which the values of all parameters, or corrections which depend on other species or the matrix, are known or can be calculated with approximate uncertainty.*

The most important analytical methods considered by CCQM in this context are:

- isotope dilution mass spectrometry (IDMS),
- gravimetry,
- tritrimetry,
- coulometry.

CCQM has also defined the new term "primary reference material":

"A primary reference material is one having the highest metrological qualities and whose value is determined by means of a primary method".

Well known cases are the application of gravimetry (for gas mixtures and isotopic CRMs) and thermal ionisation IDMS (for nuclear CRMs).

3.1.3   Types of Certifiers

Reference materials can be certified by international, national (governmental) or private institutions, with the weight of the three types depending on legislation in a specific country.

International Certifiers

The European BCR System and Its Specific Rules. For the past 25 years the European Community has funded the production and certification of reference materials for the European market with the help of an institution called the BCR (Bureau Communautaire de Référence). The BCR, which is now part of the European Community's Standards, Measurements and Testing Programme, acts as a certifier by:

- calling for proposals of CRM certification projects;
- evaluating the proposals;
- granting funds for the proposals found to be worthwhile;
- initiating the formation of consortiums of participating laboratories, one of which has to act as a project coordinator;
- evaluating and checking the results of the certification procedure presented by the coordinator through a certification committee;
- accepting the responsibility for taking the CRMs in stock, for marketing them and for post-certification stability testing.

Work for a BCR reference material is based on a set of guidelines [3.10]
which describe in detail all steps of the production and certification procedure
from the beginning of the planning phase, through the statistical tools for
data evaluation to the contents of the certification report.

Although the BCR guidelines allow different strategies to be applied for
certification, the recommended one is the interlaboratory cooperation approach described above (with about 15 participants). As the guidelines say:

*"The use of widely different principles or methods, in different laboratories,
often with fully independent calibration, leads indeed to an extremely low probability that all labs would make the same systematic error. If then their results
agree within random experimental error, this means that the common result
is accurate to within an uncertainty which can be calculated by statistics".*

The BCR guidelines ask for the qualification of new participants to be
demonstrated either by proof of relevant experience or, preferably, by successfully taking part in a qualification experiment with a similar material of
known composition.

The results of the certification have to be carefully described in a certification report which, so far, is distributed together with the certificate with
each unit of the CRM.

BCR reference materials are the property of the European Community.
The income from CRM sales is used to fund the production and certification
of new batches of materials when stocks are depleted.

The Institute for Reference Materials and Measurements (IRMM).
IRMM, Retieseweg, B-2440 Geel, Belgium, is one of the joint research centers
of the European Community, and is closely related to the BCR system. It is
in charge of keeping the stock of BCR CRMs, checking their quality during
storage, replenishing exhausted CRM batches and distributing the materials
to consumers or sub-distributors.

Apart from this, IRMM runs its own programme of certification mainly
in the field of radioactive and stable isotopic CRMs. Due to the fact that
mass spectrometric measurements and isotope dilution analyses are run by
IRMM at a very high level of quality, these CRMs (some of which fulfill the
requirements set for primary CRMs) enjoy an excellent reputation.

National Reference Material Certifiers

Most developed countries produce and certify CRMs to some extent by their
national reference institutes or under their coordination and supervision. The
importance of these CRMs in the national context depends on the role the
national institute was given by the legislator.

The US National Institute of Standards and Technology (NIST).
NIST is worldwide the most important certifier of CRMs. Its reputation is

not only based on many years of experience in a wide variety of fields of work, but on its definition by US legislation as the provider of reference materials to which American industry is compelled to refer.

So far NIST has almost exclusively used the certification strategies described above as "elite group of methods approach" and "primary method approach". Furthermore, all certifications are based on measurement data produced by NIST laboratories. Among the many methods used, neutron activation analysis should be mentioned as a method yielding results highly independent from those of other methods. It was developed by NIST for internal use and is now a reliable and frequently used analytical method.

NIST has created the term Standard Reference Material (SRM) as a trade mark for its certified reference materials. Private American reference material producers who refer to SRMs in the process of their own RM production, are entitled to claim a "traceable to NIST" quality aspect for their products, if certain requirements, set down by NIST with respect to their quality management system, are fulfilled.

Recently NIST has started to enter a system of mutual CRM recognition, based on experimental comparison of NIST CRMs with CRMs of other national standard institutes. This would allow access to the American market for foreign products in fields of application regulated by legislation.

A Selection of Other Important National Reference Material Certifiers

China. National Research Center for Certified Reference Materials (NC-CRM), No.7 District 11, Hepingjie, Chaoyangqu, Beijing, 100013, China.

This institution currently offers over 260 certified reference materials, the certified values of which are in many cases based on the application of primary analytical methods.

France. Laboratoire National d'Essais (LNE), 1 rue Gaston Boissier, 75724 Paris, Cedex 15, France.

Germany. Bundesanstalt für Materialsforschung und -prüfung (BAM), Unter den Eichen 87, 12205 Berlin, Germany.

BAM has a long tradition in the certification of metal alloy and gas mixture CRMs, but has recently started to deal with the certification of environmental materials and porous substances, too. In addition, BAM runs a program for the definition of primary reference materials in the 99.99% and 99.9999% purity classes.

The certification of BAM reference materials follows a set of guidelines which are an adjusted version of the BCR guidelines.

Netherlands. Nederlands Meetinstituut (NMI), P.O. Box 654, 2600 AR Delft, The Netherlands.

NMI has achieved international recognition for its gas mixture CRMs.

United Kingdom. Laboratory of the Goverment Chemist (LGC), Queens Road, Teddington, Middlesex TW11 0LY, United Kingdom. National Physical Laboratory (NPL), Teddington, Middlesex TW11 0LW, United Kingdom

LGC is a privatized national reference institute, offering a wide variety of reference materials. NPL is in charge of producing and keeping the British gas mixture standards.

Private Certifiers

Private firms or institutions often act as certifiers in specific fields, where their general activities have resulted in the the accumulation of sufficient experience and expertise (e.g. steel firms producing steel CRMs or gas suppliers producing gas mixture CRMs). Also, branches of a national industry can establish private institutes for the production and certification of reference materials in their respective fields of interest. Other than these two groups of private certifiers, the British firm Bureau of Analysed Samples Ltd., Middlesbrough, Cleveland TS8 9EA, certifies and offers a wide range of different CRMs.

3.2   Definitions of Terms and Modes Used at NIST for Value-Assignment of Reference Materials for Chemical Measurement

W. May, R. Parris, C. Beck, J. Fassett, R. Greenberg, F. Guenther, G. Kramer, S. Wise and T. Gills, J. Colbert, R. Gettings, B. MacDonald

Standard Reference Materials® (SRMs®) are certified reference materials (CRMs) issued under the National Institute of Standards and Technology (NIST) trademark that are well characterized by using state-of-the-art measurement methods and/or technologies for the determination of chemical composition and/or physical properties. Traditionally, SRMs have been the primary tools that NIST (formerly National Bureau of Standards) provides to the user community for achieving chemical measurement quality assurance and traceability to national standards.

This chapter provides definitions of the terms and descriptions of NIST's current practices for value-assigning SRMs and reference materials (RMs) used for calibrating and/or validating instrumentation and/or methods and procedures used for chemical measurements. The terms and modes as described in this document are applicable for reference materials that support chemical measurements issued by NIST as of October 1, 1998.

Table 3.1 lists the seven modes used at NIST for value-assigning SRMs and RMs for chemical measurements and links the modes to three possible data quality descriptors: NIST certified values, NIST reference values, and NIST information values. A *NIST Certified Value* represents data for which NIST has the highest confidence in its accuracy in that all known or suspected sources of bias have been fully investigated or accounted for by NIST. A *NIST Reference Value* is a best estimate of the true value provided by NIST where all known or suspected sources of bias have not been fully investigated by NIST. A *NIST Information Value* is a value that will be of interest and use to the SRM/RM user, but insufficient information is available to assess the uncertainty associated with the value. Definitions of these modes are given in Sect. 3.2.2.

Table 3.1. Modes used at NIST for value-assignment of reference materials for chemical measurements. A, NIST Certified Value; B, NIST Reference Value; C, NIST Information Value

	A	B	C
1. Certification at NIST using a single primary method with confirmation by other method(s)	×		
2. Certification at NIST using two independent critically evaluated methods	×	×	
3. Certification/value-assignment using one method at NIST and different methods by outside collaborating laboratories	×	×	
4. Value-assignment based on measurements by two or more laboratories using different methods in collaboration with NIST		×	×
5. Value-assignment based on a method-specific protocol		×	×
6. Value-assignment based on NIST measurements using a single method or measurements by an outside collaborating laboratory using a single method		×	×
7. Value-assignment based on selected data from interlaboratory studies		×	×

The choice of mode(s) to be used in the value-assignment for any SRM for chemical measurements is based on our previous experience and knowledge of the specific matrix, analyte(s) of interest, current measurement capabilities, the quality of the analytical methods results, and the intended use of the material.

The final designation of an assigned-value for an SRM as a NIST certified value, NIST reference value, or NIST information value is based on the specific value-assignment mode used and the assessed quality of the resulting data relative to the intended use of the material.

3.2.1 NIST Practices for Value-Assignment of SRMs and RMs for Chemical Measurements

Generally, NIST does not make or fabricate the materials from which SRMs are produced. Rather, U.S. industry, scientific groups, or companies on contract to NIST provide materials that meet NIST specifications.

Techniques and methods used at NIST for providing certified values for SRMs for chemical measurements are critically evaluated and have demonstrated accuracy in the matrix under investigation. Potential sources of error for such methods are evaluated and addressed [3.11,12]. Methods that are "ratio-based" (i.e., that require instrumental comparison versus calibrants of a known quantity of the measurand) use high-purity, well characterized primary reference compounds or species as their basis for calibration (either directly or through gravimetrically prepared calibration solutions, e.g., NIST elemental solution SRMs).

The details of NIST methods and their testing are well documented (i.e., internal NIST Reports of Analysis) and often published in refereed technical journals. When results from outside laboratories are used in the value-assignment process, the NIST Chemical Science and Technology Laboratory (CSTL) is responsible for the selection of the laboratories and the technical evaluation of these reported data.

Appropriate control materials are concurrently analyzed in all value-assignment activities, both within NIST and by any outside collaborating laboratories. When available, appropriate SRMs or CRMs from other national metrology laboratories are used for this purpose.

3.2.2 Modes for Value-Assignment and/or Certification

The quality of assigned values for any CRM is based on the existence and application of sound measurement principles and practices. It is with this basic premise that we provide the following seven modes used at NIST to acquire analytical data for the value assignment of our SRMs and RMs for chemical measurements and link these modes to three data quality descriptors: NIST Certified Values, NIST Reference Values, and NIST Information Values.

Certification at NIST Using a Single Primary Method with Confirmation by Other Method(s). Mode 1

The Consultative Committee on the Quantity of Material (CCQM) [3.13] has described a primary method as:

"A primary method of measurement is a method having the highest metrological properties, whose operation can be completely described and understood, for which a complete uncertainty statement can be written down in terms of SI units.

A primary direct method: measures the value of an unknown without reference to a standard of the same quantity.

A primary ratio method: measures the value of a ratio of an unknown to a standard of the same quantity; its operation must be completely described by a measurement equation."

Certification at NIST using a single primary method is only possible when (with the exception of special cases noted in a–c below [3.14]):

- All potentially significant sources of error have been evaluated explicitly for the application of the method and the matrix under investigation; a short written description is provided in the Report of Analysis for other sources of error that might reasonably be present and why they are not expected to be significant in this particular case.
- Confirmation of measurements by a primary NIST method can be accomplished by one or more of the following:
  - determination of certified constituents in other SRM(s) or CRM(s) of similar matrix and constituent concentration range;
  - a second NIST technique with appropriate controls;
  - results of measurements from selected outside collaborating laboratories with appropriate experience.

The required level of agreement between the primary method and any confirmatory method(s) must be predetermined and specified in the experimental plan.

a) Certification of Gaseous Mixtures at NIST Using a Primary Method. Certification of gaseous mixture SRMs at NIST requires the following:

- Primary standard suites must be prepared gravimetrically from well characterized starting materials and demonstrated to be internally consistent by a well characterized analytical method.
- NIST primary standards must be intercompared with primary standards from other National Metrology Laboratories or verified by a second NIST independent technique.
- Primary standards must be documented to be stable for a minimum of two years.
- SRMs must be value-assigned relative to the NIST primary standard suites.
- Uncertainty associated with the certified value must include contributions from the uncertainties associated with the primary standard suite, the analytical ratio method used to compare the SRM and primary standards, and the heterogeneity of the SRM lot.
- The lot homogeneity must be determined by NIST analysis of all samples.
- Absence of significant impurities in the SRMs must be verified by NIST analyses.

b) Certification of NIST pH SRMs. Certification of NIST pH SRMs requires the following:

- Homogeneity of the candidate material(s) for each pH SRM must be evaluated by intercomparisons of randomly selected aliquots of candidate material(s) normalized to the preceding issuance of the corresponding pH SRM using a glass electrode.
- As an internal control measure, a candidate material must be rejected if a significant difference is observed between the mean pH value of the current candidate material and the certified pH of the previous SRM issue (unless redetermination of the pH of the previous issue of the SRM using a Harned cell indicates a significant change from its certified pH value).
- Certification of each pH SRM must be performed using cells without liquid junction (Harned cells) at each temperature of interest using at least three independently prepared buffer solutions of composite samples of the candidate SRM.
- Uncertainty associated with the certified value must include the measurement uncertainty (in potential of Harned cell), the uncertainty in standard potential of Ag, AgCl reference electrodes, the theoretical uncertainty in the conventional calculation of -log γ_{Cl}, and the replication uncertainty for the overall pH value-assignment.

c) Certification of NIST Optical Filter SRMs. Specific requirements for the certification of NIST optical filter SRMs include:

Photometry

- Regular transmittance scale is maintained on the National Reference Spectrophotometer in the NIST Analytical Chemistry Division which is validated by the double-aperture method of light addition and benchmarked through international intercomparisons using optical filter artifact standards.
- Solid (neutral glass and metal-on-silica) SRMs are assigned certified values for transmittance and/or transmittance density at specified wavelengths by individual measurement of each artifact on the National Reference Spectrophotometer.
- Liquid or powder SRMs are assigned certified values for absorbance per unit pathlength or specific absorptivity at specified wavelengths by batch certification on the National Reference Spectrophotometer using a random sampling from the batch.
- The uncertainty for each assigned photometric value includes components to account for the precision and accuracy of the instrument; heterogeneity, temporal drift, and thermal characteristics of the artifact; and the geometry of sample positioning. Uncertainties are not individually evaluated but are based on pooled measurements with more than 30 degrees of freedom. Uncertainties are reassessed annually for continuously produced solid standards or with each reissue for batch-certified standards.

- A control filter is run with all data acquisitions, and the data are used to "control-chart" the measurement process and verify consistent performance.

Wavelength

- SRMs are assigned certified values for peak wavelength or wavenumber by comparison to atomic wavelengths (ultimately traceable to the standard meter) using a transfer spectrometer.
- Wavelength standards are batch certified using a random sampling from the lot. The uncertainty for each peak position in a wavelength standard includes components to account for the calibration accuracy of the transfer spectrometer, the precision in locating the standard peaks, and relevant temperature coefficients over the specified temperature range of valid certification.

Certification at NIST Using Two Independent Critically Evaluated Methods. Mode 2

A second mode of certification for NIST SRMs involves the use of two or more critically evaluated independent methods [3.15,16]. Method independence is of critical importance, and while it is rare that two analytical methods have completely different sources of error and variability, they are chosen so that the most significant sources of error are different. For example, the following considerations are carefully evaluated:

- Methods are selected to minimize common steps in sample preparation and the final analytical measurement techniques.
- Methods rely on different physical, spectroscopic, or chemical phenomena that generate the analytical response.
- Methods/procedures selected are appropriate for the required precision and accuracy for measurement of the analyte(s) of interest in the matrix.
- The criteria for between-method agreement required for certification is predetermined and documented in the experimental plan.

Certification/Value-Assignment Using One Method at NIST and Different Methods by Outside Collaborating Laboratories. Mode 3

In some cases, there does not exist a suitable second independent method at NIST. For these instances, we carefully select outside laboratories to collaborate on the certification process. Ideally this collaboration begins at the very start of the experimental design process. In this way, both NIST and outside laboratory analysts are able to coordinate the details of the measurement, data analysis, and reporting requirements for the SRM with careful attention to the following:

- The NIST method and the outside collaborating laboratory's methods must have been critically evaluated and demonstrated to provide accurate results for the matrix under investigation.
- The method(s) used by outside collaborating laboratories should be different from the method used at NIST as required by the "two independent NIST methods" mode (see mode 2).
- Data reporting requirements for outside collaborating laboratories should be specified in the experimental plan and reports should contain sufficient information to evaluate all significant sources of uncertainty.

Value-Assignment Based on Measurements by Two or More Laboratories Using Different Methods in Collaboration with NIST. Mode 4

This mode can be used to provide NIST reference values or NIST information values for an SRM, e.g., in instances in which there do not exist suitable methods at NIST. This mode requires the following:

- The outside collaborating laboratory methods must have demonstrated accuracy in the matrix under investigation.
- Analyses provided by the outside collaborating laboratories should involve at least two different methods (see mode 2).
- Data reporting requirements for the outside collaborating laboratories are specified in the experimental plan and their report should contain sufficient information to evaluate all significant sources of uncertainty, unless a large number of labs/methods submit data, in which case the "interlaboratory study" criteria apply (see mode 7).

Value-Assignment Based on a Method-Specific Protocol. Mode 5

In cases of method-defined parameters, the value of the parameters of interest result from the appropriate and validated use of a defined protocol. Appropriate implementation of this mode requires the following:

- The protocol used should be one that is recognized by the user community as the prescribed method for measurement of the analyte (or property) of interest in this matrix.
- Only data from experienced practitioners of the protocol should be used.
- Measurements using the method-specific protocol can be made by NIST, outside laboratories, or both.
- Method-specific value-assignment should typically involve no fewer than three experienced practitioners of the method.

Value-Assignment Based on NIST Measurements Using a Single Method or Measurements by an Outside Collaborating Laboratory Using a Single Method. Mode 6

In some cases the intended use by the measurement community does not require a NIST certified value as an assigned value. This mode can be used to provide NIST reference values or NIST information values.

- The NIST method used is typically one that would be used in the "two independent NIST methods" mode (see mode 2), i.e., the method may have been used in the past as one of several methods for SRM certification, but in this instance was the only method used.
- Method used by outside laboratory must have been demonstrated to provide appropriate precision and accuracy in the matrix under investigation.
- Data reporting requirements for outside collaborating laboratories are specified in the experimental plan and their reports should contain sufficient information to permit evaluation of significant sources of uncertainty.

Value-Assignment Based on Selected Data from Interlaboratory Studies. Mode 7

This mode allows NIST to take advantage of interlaboratory studies designed for purposes other than value-assignment of reference materials. In this mode:

- The particular study must be well documented and organized by a reputable organization.
- NIST Chemical Science and Technology Laboratory (CSTL) is responsible for evaluating the appropriateness of analytical procedures to identify a subset of results to be selected for use in value-assignment.

3.2.3   Definition of Terms

NIST references a number of definitions in connection with the production, certification, and use of its SRMs and RMs. The uses of the terms "certified values", "reference values", etc. have multiple meanings based on the intent and practices of a particular reference material supplier. Certain definitions, adopted for NIST use, are derived from international guides and standards on reference materials and measurements while others have been developed by NIST to describe those activities unique to NIST operations and philosophy. To avoid any ambiguity, this publication provides definitions of the terms as they are currently used by NIST and a description of NIST's current practices for value-assigning SRMs and RMs that support chemical measurements. A listing of NIST-adopted and NIST-developed definitions follows.

Reference Material (RM). Material or substance one or more of whose property values are sufficiently homogeneous and well established to be used for the calibration of an apparatus, the assessment of a measurement method, or for assigning values to materials ([3.17], Sect. 6.13).

Certified Reference Material (CRM). Reference material, accompanied by a certificate, one or more of whose property values are certified by a procedure which establishes traceability to an accurate realization of the unit in which the property values are expressed, and for which each certified value is accompanied by an uncertainty at a stated level of confidence ([3.17], Sect. 6.14).

NIST Standard Reference Material® (SRM®). A CRM issued by NIST that also meets additional NIST-specified certification criteria. NIST SRMs are issued with certificates of analysis or certificates that report the results of their characterizations and provide information regarding the appropriate use(s) of the material.

NIST Traceable Reference Material™ (NTRM™). A commercially-produced reference material with a well-defined traceability linkage to existing NIST standards for chemical measurements. This traceability linkage is established via criteria and protocols defined by NIST to meet the needs of the metrological community to be served.

NIST Certified Value. A value reported on an SRM certificate/certificate of analysis for which NIST has the highest confidence in its accuracy in that all known or suspected sources of bias have been fully investigated or accounted for by NIST. Values are generally referred to as certified when modes 1, 2, or 3 have been used for value-assignment and all the criteria for that mode are fulfilled. These three modes all require NIST measurements and oversight of the experimental design for the value-assignment process. The uncertainty associated with a certified value generally specifies a range within which the true value is expected to lie at a level of confidence of approximately 95% if the sample is homogeneous. If significant sample heterogeneity is included, the uncertainty generally represents a prediction interval within which the true values of 95% of all samples are expected to lie at a stated level of confidence.

Uncertainty of a Certified Value. An estimate attached to a certified value of a quantity which characterizes the range of values within which the "true value" is asserted to lie with a stated level of confidence ([3.18], Sect. 3.4). Uncertainty of a measurement: parameter associated with the result of a measurement that characterizes the dispersion of the values that could reasonably be attributed to the measurand ([3.17], Sect. 3.9).

NIST Reference Value (Formerly Called Noncertified Value) for Chemical Composition and Related Properties. A NIST reference value is a best estimate of the true value provided on a NIST certificate/certificate of analysis/report of investigation where all known or suspected sources of bias have not been fully investigated by NIST. Reference values are generally determined using the following modes:

- Modes 4, 5, or 6 are used.
- Results from modes 4, 5, 6, or 7 are used in which the intended use of the value by the measurement community does not require that it be a certified value.
- Mode 2 or 3 is used but there is lack of sufficient agreement among the multiple methods.
- Mode 7 can be used in special cases, e.g., when results are obtained from another national metrology institute with whom NIST has historical comparability data for the method(s) used for the specific matrix/analyte combination.

The uncertainty associated with a NIST reference value may not include all sources of uncertainty and may represent only a measure of the precision of the measurement method(s).

NIST Information Value. A NIST information value is considered to be a value that will be of interest and use to the SRM/RM user, but insufficient information is available to assess the uncertainty associated with the value. Typically, the information value has no reported uncertainty listed on the certificate and has been derived from one of the following value-assignment modes:

- Results from modes 4, 5, 6, or 7 in which the intended use of the value by the measurement community does not require that it be a certified or reference value (e.g., information about the composition of the matrix such as the value of "total organic carbon" of a sediment material may be useful to the user in selecting an appropriate analytical method).
- The results from modes 4, 5, 6, or 7 lack sufficient information to assess the uncertainty.
- Results are provided from outside NIST as supplemental information on the SRM matrix and are not measurements typically made at NIST but may be of interest to the user.

3.2.4   Appendix: Standard Reference Material® 1944. New York/New Jersey Waterway Sediment

Standard Reference Material (SRM) 1944 is a mixture of marine sediment collected near urban areas in New York and New Jersey. SRM 1944 is intended for use in evaluating analytical methods for the determination of selected polycyclic aromatic hydrocarbons (PAHs), polychlorinated biphenyl (PCB) congeners, chlorinated pesticides, and trace elements in marine sediment and similar matrices. All of the constituents for which *certified*, *reference*, and *information* values are provided in SRM 1944 were naturally present in the sediment material before processing. Over 150 constituents or constituent elements were value-assigned in this SRM. Some examples of the three types of value-assignments are described and presented in the accompanying tables.

Certified Concentration Values. Certified values for concentrations, expressed as mass fractions, for 24 PAHs and nine trace elements are provided in Tables 3.2 and 3.3. A NIST certified value is a value for which NIST has the highest confidence in its accuracy in that all known or suspected sources of bias have been investigated or accounted for by NIST. The certified values for the PAHs, PCB congeners, and chlorinated pesticides are based on the agreement of results obtained at NIST from two or more chemically independent analytical techniques. The certified values for the trace elements are based on NIST measurements by one technique and additional results from several collaborating laboratories.

Reference Concentration Values. Reference values for concentrations, expressed as mass fractions, are provided for 32 additional PAHs (some in combination) in Table 3.4 and 19 additional inorganic constituents in Tables 3.5 and 3.6. Reference values are noncertified values that are the best estimate of the true value; however, the values do not meet the NIST criteria for certification and are provided with associated uncertainties that may reflect only measurement precision, may not include all sources of uncertainty, or may reflect a lack of sufficient statistical agreement among multiple analytical methods.

Information Concentration Values. Information values for concentrations, expressed as mass fractions, are provided in Table 3.7 for eight additional trace elements. An information value is considered to be a value that will be of interest and use to the SRM user, but insufficient information is available to assess the uncertainty associated with the value or only a limited number of analyses were performed.

Table 3.2. Certified concentrations for selected PAHs in SRM 1944

PAHs	Mass fractions (dry-mass basis)[a,b] [mg/kg]		
Naphthalene[c,d,e,f,g]	1.65	±	0.31
Phenanthrene[c,d,e,f,g]	5.27	±	0.22
Anthracene[c,d,e,f,g]	1.77	±	0.33
Fluoranthene[c,d,e,f,g]	8.92	±	0.32
Pyrene[c,d,e,f,g]	9.70	±	0.42
Benzo[c]phenathrene[c,d,e,f,h]	0.76	±	0.10
Benz[a]anthracene[c,d,e,f,g,h]	4.72	±	0.11
Chrysene[h,k]	4.86	±	0.10[i]
Triphenylene[h,k]	1.04	±	0.27
Benzo[b]fluoranthene[g,h,j]	3.87	±	0.42
Benzo[j]fluoranthene[h,j]	2.09	±	0.44
Benzo[k]fluoranthene[c,d,e,f,g,h,j]	2.30	±	0.20
Benzo[a]fluoranthene[c,d,e,f,h,,j]	0.78	±	0.12
Benzo[e]pyrene[c,d,e,f,h,j]	3.28	±	0.11
Benzo[a]pyrene[c,d,e,f,g,h,j]	4.30	±	0.13
Perylene[c,d,e,f,g,h,j]	1.17	±	0.24
Benzo[ghi]perylene[c,d,e,f,j,k]	2.84	±	0.10
Indeno[1,2,3-cd]pyrene[c,d,e,f,j,k]	2.78	±	0.10
Dibenz[a,j]anthracene[c,d,e,f,j,k]	0.500	±	0.044
Dibenz[a,c]anthracene[j,k]	0.335	±	0.013
Dibenz[a,h]anthracene[j,k]	0.424	±	0.069
Pentaphene[c,d,e,f,j,k]	0.288	±	0.026
Benzo[b]chrysene[c,d,e,f,j,k,h]	0.63	±	0.10
Picene[c,d,e,f,j,k]	0.518	±	0.093

[a] Concentrations reported on dry-mass basis; material as received contains approximately 1.3% moisture.

[b] The results are expressed as the certified value ± the expanded uncertainty. Each certified value is a mean of the means from two or more analytical methods, weighted as described in Paule and Mandel [3.20]. Each uncertainty, computed according to the CIPM approach as described in the ISO guide [3.21], is an expanded uncertainty at the 95% level of confidence, which includes random sources of uncertainty within each analytical method as well as uncertainty due to the drying study. The expanded uncertainty defines a range of values within which the true value is believed to lie, at a level of confidence of approximately 95%.

[c] GC/MS (I) on 5% phenyl-substituted methylpolysiloxane phase after Soxhlet extraction with DCM.

[d] GC/MS (II) on 5% phenyl-substituted methylpolysiloxane phase after Soxhlet extraction with DCM.

[e] GC/MS (III) on 5% phenyl-substituted methylpolysiloxane phase after Soxhlet extraction with 50% hexane/50% acetone.

[f] GC/MS (IV) on 5% phenyl-substituted methylpolysiloxane phase after PFE with 50% hexane/50% acetone.

[g] LC-FL of total PAH fraction after Soxhlet extraction with 50% hexane/50% acetone.

[h] GC/MS (Sm) using a smectic liquid crystalline phase after Soxhlet extraction with DCM.

Table 3.2. (Continued)

[i] The uncertainty interval for chrysene was widened based on expert consideration of the analytical methods and analysis of the data for all PAHs, which suggests that the half-widths of the expanded uncertainties should not be less than 2%.
[j] GC/MS (V) on 50% phenyl-substituted methylpolysiloxane phase of extracts from GC/MS (III) and GC/MS (IV).
[k] LC-FL of isomeric PAH fractions after Soxhlet extraction with 50% hexane/50% acetone.

Table 3.3. Certified concentrations for selected inorganic constituents in SRM 1944

Elements	Degrees of freedom	Mass fractions (dry-mass basis)[a,b] [%]
Aluminum[c,d,e]	4	5.33 ± 0.49
Iron[c,d,e]	6	3.53 ± 0.16

		Mass fractions (dry-mass basis)[a,b] [mg/kg]
Arsenic[c,d,e,f,g]	10	18.9 ± 2.8
Cadmium[c,f,h,i]	6	8.8 ± 1.4
Chromium[c,d,f,g,i]	9	266 ± 24
Lead[c,h,i]	5	330 ± 48
Manganese[c,d,e]	8	505 ± 25
Nickel[c,g,h,i]	6	76.1 ± 5.6
Zinc[c,d,e,g,i]	9	656 ± 75

[a] The results are expressed as the certified value ± the expanded uncertainty. The certified value is the mean of four results: (1) the mean of NIST INAA or ID-ICPMS analyses, (2) the mean of two methods performed at NRCC, and (3) the mean of results from seven selected laboratories participating in the NRCC intercomparison exercise, and (4) the mean results from INAA analyses at IAEA. The expanded uncertainty in the certified value is equal to $U = ku_c$, where u_c is the combined standard uncertainty and k is the coverage factor, both calculated according to the ISO Guide [3.21]. The value of uc is intended to represent at the level of one standard deviation the combined effect of all the uncertainties in the certified value. Here uc accounts for both possible method biases, within-method variation, and material inhomogeneity. The coverage factor, k, is the Student's t-value for a 95% prediction interval with the corresponding degrees of freedom. Because of the material inhomogeneity, the variability among the measurements of multiple samples can be expected to be greater than that due to measurement variability alone.
[b] Concentrations reported on dry-mass basis; material as received contains approximately 1.3% moisture.
[c] Results from five to seven laboratories participating in the NRCC interlaboratory comparison exercise.
[d] Measured at NIST using INAA.
[e] Measured at NRCC using ICPAES.
[f] Measured at NRCC using GFAAS.
[g] Measured at IAEA using INAA.
[h] Measured at NIST using ID-ICPMS.
[i] Measured at NRCC using ID-ICPMS.

Table 3.4. Reference concentrations for selected PAHs in SRM 1944

PAHs	Mass fractions (dry-mass basis)[a,b] [mg/kg]		
1-Methylnaphthalene[c,d,e,f]	0.52	±	0.03
2-Methylnaphthalene[c,d,e,f]	0.95	±	0.05
Biphenyl[c,d,e,f]	0.32	±	0.07
Acenaphthene[c,d,e,f]	0.57	±	0.03
Fluorene[c,d,e,f]	0.85	±	0.03
Dibenzothiophene[d,e,f]	0.62	±	0.01[g]
1-Methylphenanthrene[c,d,e,f]	1.7	±	0.1
2-Methylphenanthrene[c,d,e,f]	1.90	±	0.06
3-Methylphenanthrene[c,d,e,f]	2.1	±	0.1
4-Methylphenanthrene and 9-Methylphenanthrene[c,d,e,f]	1.6	±	0.2
2-Methylanthracene[c,d,e,f]	0.58	±	0.04
3,5-Dimethylphenanthrene[c]	1.31	±	0.04
2,6-Dimethylphenanthrene[c]	0.79	±	0.02[g]
2,7-Dimethylphenanthrene[c]	0.67	±	0.02[g]
3,9-Dimethylphenanthrene[c]	2.42	±	0.05[g]
1,6-, 2,9-, and 2,5-Dimethylphenanthrene[c]	1.67	±	0.03[g]
1,7-Dimethylphenanthrene[c]	0.62	±	0.02[g]
1,9- and 4,9-Dimethylphenanthrene[c]	1.20	±	0.03[g]
1,8-Dimethylphenanthrene[c]	0.24	±	0.01[g]
1,2-Dimethylphenanthrene[c]	0.28	±	0.01[g]
8-Methylfluoranthene[c]	0.86	±	0.02[g]
7-Methylfluoranthene[c]	0.69	±	0.02
1-Methylfluoranthene[c]	0.66	±	0.02[g]
3-Methylfluoranthene[c]	2.46	±	0.07
2-Methylpyrene[c]	1.81	±	0.04[g]
4-Methylpyrene[c]	1.44	±	0.03[g]
1-Methylpyrene[c]	1.29	±	0.03
Anthanthrene[h]	0.9	±	0.1

[a] Concentrations reported on dry-mass basis; material as received contains approximately 1.3% moisture.

[b] The reference value for each analyte is the equally-weighted mean of the means from two or more analytical methods or the mean from one analytical technique. The uncertainty in the reference value defines a range of values that is intended to function as an interval that contains the true value at a level of confidence of 95%. This uncertainty includes sources of uncertainty within each analytical method, among methods, and from the drying study.

[c] GC/MS (I) on 5% phenyl-substituted methylpolysiloxane phase after Soxhlet extraction with DCM.

[d] GC/MS (II) on 5% phenyl-substituted methylpolysiloxane phase after Soxhlet extraction with DCM.

[e] GC/MS (III) on 5% phenyl-substituted methylpolysiloxane phase after Soxhlet extraction with 50% hexane/50% acetone.

Table 3.4. (Continued)

---

[f] GC/MS (IV) on 5% phenyl-substituted methylpolysiloxane phase after PFE with 50% hexane/50% acetone.

[g] The uncertainty interval for this compound was widened in accordance with expert consideration of the analytical procedures, along with the analysis of the data as a whole, which suggests that the half-widths of the expanded uncertainties should not be less than 2%.

[h] LC-FL of isomeric PAH fractions after Soxhlet extraction with 50% hexane/50% acetone.

Table 3.5. Reference concentrations for selected inorganic constituents in SRM 1944

Elements	Degrees of freedom	Mass fraction (dry-mass basis)[a,b] [%]		
Silicon[c,d]	81	31	±	3
		Mass fraction (dry-mass basis)[a,b] [mg/kg]		
Beryllium[c,h]	17	1.6	±	0.3
Copper[c,d,f]	101	380	±	40
Mercury[c,i]	18	3.4	±	0.5
Selenium[c,e,f]	24	1.4	±	0.2
Silver[c,d,e,g]	8	6.4	±	1.7
Thallium[c,f]	12	0.59	±	0.1
Tin[c,f]	22	42	±	6

[a] The results are expressed as the reference value ± the expanded uncertainty. The reference value is the equally weighted mean of available results from: (1) NIST INAA analyses, (2) two methods performed at NRCC, (3) results from seven selected laboratories participating in the NRCC intercomparison exercise, and (4) results from INAA analyses at IAEA. The expanded uncertainty in the reference value is equal to $U = ku_c$ where u_c is the combined standard uncertainty and k is the coverage factor, both calculated according to the ISO Guide [3.21]. The value of u_c is intended to represent at the level of one standard deviation, the uncertainty in the value. Here uc accounts for both possible method differences, within-method variation, and material inhomogeneity. The coverage factor, k, is the Student's t-value for a 95% prediction interval with the corresponding degrees of freedom. Because of material inhomogeneity, the variability among the measurements of multiple samples can be expected to be greater than that due to measurement variability alone.
[b] Concentrations reported on dry-mass basis; material as received contains approximately 1.3% moisture.
[c] Results from five to seven laboratories participating in the NRCC interlaboratory comparison exercise.
[d] Measured at NRCC using GFAAS.   [e] Measured at NIST using INAA.
[f] Measured at NRCC using ID-ICPMS. [g] Measured at IAEA using INAA.
[h] Measured at NRCC using ICPAES.   [i] Measured at NRCC using CVAAS.

Table 3.6. Reference concentrations for selected inorganic constituents in SRM 1944 as determined by INAA

Element	Effective degrees of freedom	Mass fraction in percent (dry-mass basis)[a,b] [%]		
Calcium	21	1.0	±	0.1
Chlorine	21	1.4	±	0.2
Potassium	21	1.6	±	0.2
Sodium	25	1.9	±	0.1
		Mass fraction in mg/kg (dry-mass basis)[a,b] [mg/kg]		
Bromine	10	86	±	10
Cesium	11	3.0	±	0.3
Cobalt	10	14	±	2
Rubidium	14	75	±	2
Scandium	37	10.2	±	0.2
Titanium	21	4300	±	300
Vanadium	21	100	±	9

[a] The results are expressed as the reference value ± the expanded uncertainty. The reference value is based on the results from an INAA study. The associated uncertainty accounts for both random and systematic effects, but because only one method was used, unrecognized bias may exist for some analytes in this matrix. The expanded uncertainty in the reference value is equal to $U = ku_c$, where u_c is the combined standard uncertainty and k is the coverage factor, both calculated according to the ISO Guide [2]. The value of u_c is intended to represent at the level of one standard deviation, the uncertainty in the value. Here uc accounts for within-method variation and material inhomogeneity. The coverage factor, k, is the Student's t-value for a 95% prediction interval with the corresponding degrees of freedom. Because of material inhomogeneity, the variability among the measurements of multiple samples can be expected to be greater than that due to measurement variability alone.

[b] Concentrations reported on dry-mass basis; material as received contains approximately 1.3% moisture.

Table 3.7. Information values for concentrations for selected inorganic constituents in SRM 1944 as determined by INAA

Elements	Mass fractions (dry-mass basis)[a] [%]
Magnesium[b]	1.0
	Mass fractions (dry-mass basis)[a] [mg/kg]
Antimony[b,c]	5
Cerium[c]	65
Europium[c]	1.3
Gold[c]	0.10
Lanthanum[c]	39
Thorium[c]	13
Uranium[c]	3.1

[a] Concentration is reported on a dry-mass basis; material as received contains approximately 1.3% moisture.
[b] Measured at NIST using INAA. [c] Measured at IAEA using INAA.

3.3 Data Assessment: Influence of Homogeneity and Stability on the Reliability of Certified Amounts

S. Noack

Certified reference materials are used for calibration of analytical instruments (spark emission spectrometry, XRF) and especially for controlling the accuracy of analytical results. For faultless reckoning of accuracy using CRMs the following conditions must be met:

- The composition of any sample of the certified reference material must correspond perfectly to the certified values declared in the certificate.
- This condition has to be valid for the total time of stability guarantee.
- If a result obtained by analysis of the CRM is significantly different to the certified value (the uncertainty area of the control analysis is not overlapped by the uncertainty area of the certified value), then for the user it must be a logical conclusion that the result of the control analysis involves a systematic error considering an adequate statistical probability of error. But this conclusion must be perfectly unambiguous!
- These conditions are only satisfied if the amount validated by the certified value is affected to the total material. This in turn is only met if the distribution of the element or the component in the CRM is extremly homogeneous. The mean result of a control analysis of the CRM must not be significantly different to the certified value only because of inhomogeneous properties. Furthermore the stability of a component whose amount is certified has to be guaranteed for a given time.

The question of relevance regarding magnitude of the variation is not influenced by these considerations.

Homogeneity and stability are therefore two important aspects which critically influence the reliability of the certified amounts and the quality of CRMs. We now discuss their significance.

3.3.1 The Significance of Inhomogeneities in Quality of CRMs

The reckoning of the accuracy of analytical results is carried out by comparing the results with the certified value of the CRM. An ideal reckoning is only possible if two conditions are met:

- no uncertainty in the certified value;
- high level of homogeneity in the material as a whole.

Neither condition is met in practice.

According to the "classical" planning of interlaboratory tests for certification it can be assumed that systematic errors concerning laboratories and analytical procedures are randomized. However, this is a hypothesis which is only valid for a large number of analytical methods for the interlaboratory tests which may differ in sample preparation procedures (for example NAA, ICP-OES, XRF).

This hypothesis is only conditionally valid and certified values thus involve an uncertainty due to the planning of the interlaboratory tests and, of course, the influence of staff and analytical instruments on the result of the analysis.

An essential influence on the reliability of the certified amount and an additional contribution to its uncertainty is given by degree of homogeneity in the investigated material. This can only be tested by a hypothesis test (sampling inspection), an investigation of the whole material being, of course, impossible. Because of the random error the result of the test has an uncertainty too.

Furthermore homogeneity is necessary only between the "smallest units" analyzed. This "smallest unit" can be defined for ground materials by the minimum weight. For methods, which require compact samples, other viewpoints are valid. For spark emission spectrometry the "smallest unit" is the evaporated part of the sample. For XRF the "smallest unit" is the layer thickness affecting the irradiated area from which the excited radiation is emitted. The "smallest unit" is therefore not defined unambiguously but depends on the application for which the sample will be used.

It can be seen that "classical" interlaboratory tests involve several difficulties. A large number of laboratories carries out a number of parallel analyses (for example 4 repetitions). From the received mean values of the laboratories (or if pooling is possible, from single values) a total mean and its uncertainty are calculated. The uncertainty is usually the confidence interval. But this calculated confidence interval only characterizes the mean of the batch, and not the variation in the amounts between individual samples of the total material caused by inhomogeneities. This can cause considerable errors if inhomogeneous materials are used for certification.

The "classical" confidence interval therefore has the basic flaw that it may theoretically be equal to zero because of the division by the square root of the number of means:

$$\overline{x}_1, \overline{x}_2, \overline{x}_3, \ldots \overline{x}_n \Rightarrow \overline{\overline{x}}, s \Rightarrow \frac{t \cdot s}{\sqrt{n}} t = f(\alpha, n) \,,$$

$$\overline{\overline{x}} - \frac{t \cdot s}{\sqrt{n}} \leq \mu \leq \overline{\overline{x}} + \frac{t \cdot s}{\sqrt{n}} \,.$$

In that case the confidence interval does not characterize the variation caused by inhomogeneities between different samples but is only an indication for the reliability of the total mean ("true value").

Consequently, the use of a confidence interval is inadmissible for characterizing the uncertainty of the mean if the material is not homogeneous. In that case a region of "true values" exists (and not only one "true value"!) which differ from each other depending on the degree of inhomogeneity in the material, even when the analyst has made a high number of repetitions. One has to ensure that either the material is homogeneous according to the minimum weight or the "smallest sample" or the degree of inhomogeneity has to be taken into account for calculating the uncertainty of the certified value.

The question is whether or not homogeneous materials exist at all. The alternative is that only the degree of unavoidable inhomogeneities characterizes whether the material is usable for certification or not. Therefore in the BCR guidelines for the production and certification of BCR reference materials it will be proposed that a tolerance interval be used as uncertainty and not the confidence interval if inhomogeneities cannot be excluded.

The tolerance interval characterizes a given part of the population of *all* samples with a defined probability (the real amount of all single samples used by an analyst) and does not only characterize the mean amount of the total material. The tolerance interval can generally be used, if a small inhomogeneity is detected or has to be assumed (for example as a segregation effect at producing metallic alloys as raw material for a CRM) but cannot be detected because of a too small power of the statistical test for homogeneity. That means that the uncertainty must include the random error of the test even if an inhomogeneity cannot be significantly detected.

This can be particularly important if the precision of the procedure for testing the homogeneity is not better than the variation caused by the inhomogeneity of the material. Hence, the criterion determining success of the test is the variation of the testing procedure: the smaller the quotient of the precision of the testing procedure and the variation caused by inhomogeneities the more detectable the inhomogeneities become.

An additional aspect is that in many cases the precision of the analytical procedure cannot be determined. This is only possible if the same part of the sample can be analyzed several times, and this in turn is only possible if one can use non destructive analytical procedures like XRF. Another possibility would be the existence of a completely homogeneous sample, but such cannot be produced in practice. In all other cases the precision is the sum of the contribution to the uncertainty caused by inhomogeneity and the precision of the analytical method.

The question is whether the uncertainty of the certified value characterizes only the variation of the interlaboratory test or whether one has to suppose that the determined precision is caused by a distribution of the component in the reference material or at least a sum of the two influences.

If an inhomogeneity cannot be excluded or the degree of inhomogeneity is higher than the variation of the interlaboratory-test, then the uncertainty characterizes the distribution of the component and the mean is only a statistical value arising from to the practice. In this case it is not advisable to use the concept of a "true value"or a "true amount".

The calculation of tolerance intervals is described by Pauwels, Lamberty und Schimmel [3.19].

3.3.2 The Influence of the Stability on the Quality of CRMs

The stability of the components of a CRM has an essential influence on the reliability of a certified value as well as the homogeneity. But this influence can be reckoned only in the context of the uncertainty of the certified value.

The variation of the real amount of a CRM component caused by instability depends on time. This effect will be expressed either as a decomposition of the component (organic compounds) or by reaction with the environment of the sample (oxygen, carbon dioxide, humidity). Concerning the most frequent case which is a loss of the analyte and not considering cases involving a time dependent increase in the amount (loss of solvent, contamination by O, C or H), the following conclusions can be drawn.

There is no causal connection between the relative contribution to the uncertainty caused by the reproduciblity of the interlaboratory test and possible inhomogeneities and the instability. Because of this the upper confidence limit (instability not taken into account) drifts to lower values and approximates to the certified value. The lower limit of the confidence interval moves away from the certified value as shown in Fig. 3.1.

The consequence is that in the case of unstable components of a CRM or an unstable raw material one will frequently find results of a control analysis which are significantly lower than the certified value. Values significantly higher than the certified value will be less common.

Such misinterpretation is only avoidable if the time dependence of the stability is well known and the contribution of the instability to the uncertainty is taken into account at the moment of certification. Therefore during production of volatile organic or highly oxidisable substances, investigations to test

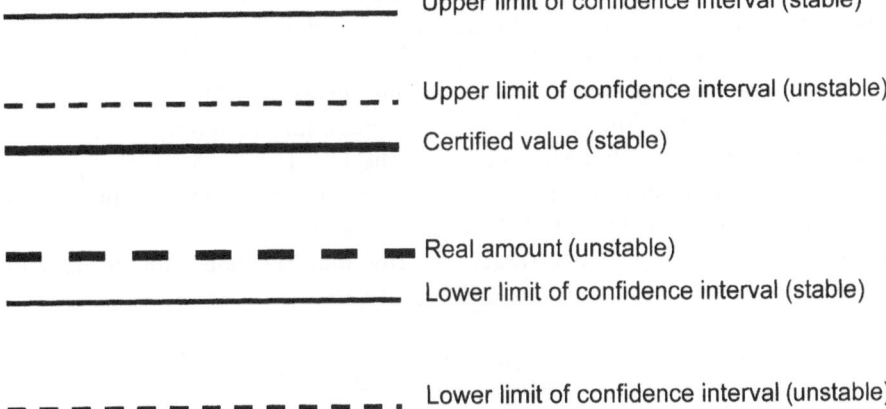

Fig. 3.1. The affect of instability on the certified value and its uncertainty

the homogeneity and the stability must be integrated into the certification procedure.

As mentioned above, for faultless use of CRMs in controlling accuracy, the composition of the CRM must in every case be in in accordance with the declaration in the certificate. For every sample used for accuracy control, calibration or matrix simulation, the real amounts must be included within the certified amount and its uncertainty. This means that for investigating influences on uncertainty, the contribution of an instability has to be taken into account as well as the contribution of inhomogeneities.

Two questions must therefore be addressed:

1. Which conditions must be met with regard to the stability of a reference material, in order that the certified value remain valid throughout the time of stability guarantee?
2. Which criteria must be considered for calculating the uncertainty of the certified value and what is the procedure for calculating to ensure that the real amounts of the components lie within the uncertainty range from the certified value.

Concerning the first question one can say that the ideal property of time stability of the certified value over many years cannot be met for many organic and/or oxidisable substances. The real problem is to determine the course of stability. For this an experiment is necessary over the time period for which the certified value should be guaranteed. This experiment can be substituted by an interpolation, if the experiment cannot be carried out over the whole time period.

Concerning the second question one can say that the variation of the certified value must not be greater than the declared uncertainty. This means that the uncertainty must include the variation of the real amount. An example is given by element solutions provided by NIST.

Naturally the influence of an instability is very small at the time of producing the CRM but increases with time. In order to avoid a varying uncertainty, the uncertainty must include the maximum contribution of the instability. As a consequence, the certified value will be corrected more than necessary early on (see Fig.3.1). The accuracy of the uncertainty thus increases with time and has its maximum at the time of maximum instability. One has the paradoxical situation that the quality of the CRM increases with time. An alternative would be to define the certified value as the mean of the amount at the beginning and the end of the time with guaranteed stability. If one has to avoid this increasing uncertainty, the raw material must not be used for certfication.

3.4   References

3.1. ISO Guide 34 (1996) Quality system guidelines for the production of reference materials

3.2. ISO Guide 31 (1981) Contents of certificates of reference materials

3.3. ISO Guide 35 (1989) Certification of reference materials – general and statistical principles

3.4. ISO/IEC Guide 65, EN 4511 (1996/1998) General requirements for bodies operating product certification systems

3.5. ISO 3534-1 (1993) Statistics – vocabulary and symbols – Part 1: Probability and general statistical terms

3.6. ISO (1993,1995) GUM: Guide to the Expression of Uncertainty in Measurement

3.7. I. Papadakis and P. de Bievre (1997) Accred. Qual. Assur. **2**, 7

3.8. ISO VIM (1993) 2nd edn. International Vocabulary of Basic and General Terms in Metrology

3.9. Pauwels J, Lamberty A and Schimmel H (1998) Accred. Qual. Assur. **3**, 180–184

3.10. Document BCR/01/97 (1997) Guidelines for the production and certification of BCR reference materials

3.11. Guidelines for Evaluating and Expressing the Uncertainty of NIST Measurement Results (1994) NIST Technical Note 1297

3.12. Guide to the Expression of Uncertainty of Measurement (1993) 1st ed. ISBN 92-67-10188-9, International Organization for Standardization (ISO)

3.13. Minutes from the Fifth Meeting (February 1998) of the Consultative Committee on the Quantity of Material (CCQM) (1998) Bureau International des Poids et Mesures (BIPM), Sèvres, France

3.14. Moody J R, Epstein M S (1991) Definitive Measurement Methods, Spectrochimica Acta, **46B**, 12

3.15. Epstein M S (1991) The Independent Method Concept for Certifying Chemical Composition Reference Materials, Spectrochimica Acta, **46B**, 12

3.16. Schiller S B and Eberhardt K B (1991) Combining Data From Independent Methods, Spectrochimica Acta, **46B**, 12

3.17. International Vocabulary of Basic and General Terms in Metrology (VIM) (1993) (2nd ed.) BIPM/IEC/IFCC/ISO/IUPAC/IUPAP/OIML, International Organization for Standardization (ISO)

3.18. Terms and Definitions Used in Connection with Reference Materials (1992) ISO Guide 30, International Organization for Standardization (ISO)

3.19. Pauwels J, Lamberty A and Schimmel H (1997) in Accred. Qual. Ass.

3.20. Paule RC, Mandel J (1982) Consensus Values and Weighting Factors, J. Research NBS **87**, 377–385

3.21. Guide to the Expression of Uncertainty in Measurement, ISBN 92-67-10188-9, 1st Ed. ISO, Switzerland, 1993; see also Taylor BN and Kuyatt CE, "Guidelines for Evaluating and Expressing Uncertainty of NIST Measurement Results," NIST Technical Note 1297, US Government Printing Office, Washington, DC (1994)

4 Reference Materials in Materials Testing

Klaus Meyer and Ralf Matschat

4.1 Strategies

Quality assurance and control is the back bone of all precision industrial and scientific activities. Industrial and technological progress can only be materialized through strict implementation of quality assurance systems. To achieve a high quality of products, reliable and accurate measurement is needed. The required quality can only be achieved or confirmed through the use of reference materials prepared and certified to a high level.

RMs in materials testing are materials or substances for which one or more properties – in most cases the concentration of a component – are sufficiently well established to be used for the calibration of an instrument (often a spectrometer), i.e. as a tool to demonstrate traceability, the assessment of a measurement, or for assigning values to materials, i.e. quality control as a tool to verify the performance of an applied method.

This article describes the aspects of production (preparation), availability and use of RMs, mainly certified for chemical composition, in the fields of metallic and inorganic non-metallic RMs concerned with materials testing. Furthermore, production (preparation) and use of RMs with certified values of particle size, surface area and pore sizes are described as examples for these particular properties. Table 4.1 shows different CRM types and selected fields of materials analysis and characterization. Categories in italics constitute the main points discussed.

4.1.1 Tasks of Materials Analysis

Maintenance of a high technical level and progress in the development and application of materials with novel or improved properties are highly dependent on the performance of analytical methods and procedures; these methods, in turn, are strongly influenced by the availability of reference materials (RMs) for the calibration of analytical instruments and for testing the trueness of analytical procedures [4.1].

The following scheme shows how the quality of an analytical result is connected with a RM:

Table 4.1. CRM types and selected categories

Chemical composition/ purity	Physical property	Special engineering property
Chemical standards	*Particle size*	Surface finish
High-purity materials	*Surface area and porosity*	Fire research
Ferrous metals	Density	Tape adhesion testing
Nonferrous metals	Thermodynamic properties	Charpy V-notch test blocks
Ceramics and glasses	Optical and electrical properties	
Semiconductors		
Inorganics		
Geological materials and ores		

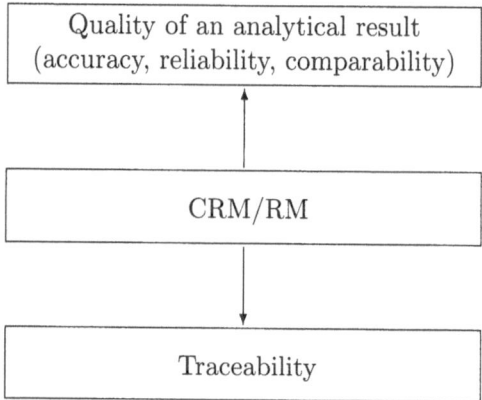

The role of certified RMs can easily be recognized. In chemical analytics, they are the links between the quality of an analytical result and its traceability, which is part of the quality of the result.

Reference materials guarantee the reliability and comparability of analytical methods in materials research and in applications. The development of new technologies and the subsequent maintenance of high quality standards in production processes involving these new technologies would be impossible without corresponding and adequate analytical methods. Due to globalization of the production and use of commodities, unified and internationally recognized evaluation rules must become the basis of quality management; this situation in turn implies a growing importance for reference materials of a high metrological level. The quality of reference materials is decisive for the quality of the analysis. By means of reference materials, one can control chemical analytical processes and directly determine the performance of analytical methods and also of the test laboratories.

Materials analysis encompasses a wide range of substances, and therefore complex analytical information is needed, comprising such diverse items as elemental and phase composition, structural analysis in some cases including the bonding state of individual elements (speciation analysis), and a characterization of the macroscopic state of the product (e.g. compact, chip-shaped or disperse). For analytical process control, fast analytical methods may be needed, facilitating immediate feedback to the production process. Finally, we must also include the topochemical aspects of analytical chemistry, especially surface and interface analysis with their frequently high lateral and depth resolution. Although for a proper understanding of properties and performance of materials it is very important to determine the spatial distribution of elements and compounds at external and internal surfaces and phase boundaries (e.g. grain boundaries), in addition to their analytical determination, we will have to exclude these analytical tasks from our presentation, since they require very specific techniques and reference materials. Problems concerning the spatial distribution within samples when preparing or applying reference materials will be considered here only in connection with homogeneity tests. Through such tests, it is possible to evaluate concentration profiles caused by diffusion or segregation, i.e. inhomogeneities which are frequently the source of errors.

Because certified reference materials for the determination of elemental composition play a dominant role and are very wide spread, this contribution puts the main emphasis on the preparation and application of RMs in the classical bulk analysis of materials for routine, precision and trace analyses.

The central task of chemical analytics comprises, at one end of the concentration scale, precision analyses in the percent range for exact determinations of content, and at the other end, trace and ultratrace analyses down to the ppm and ppb range, e.g. for characterizing superpure (high purity) materials used for the production of microelectronic devices.

In addition, we will also deal with special aspects of the preparation and application of reference materials, specifically in the analysis of disperse systems, like catalysts, adsorbents, or powders (see Sects. 4.4.5 and 4.4.6). The properties of these materials are not only determined by their chemical composition, but also by their degree of dispersity, like grain size, grain size distribution, specific surface area, pore volume, pore size and pore size distribution. Production and application of reference materials with these certified properties differs significantly from the corresponding procedures for RMs with certified elemental content; due to their fractal character, properties of the former reference materials strongly depend on the method applied, while the determination of the elemental content is usually done using at least two analytical methods based on different principles.

Since development, applications and utilization of reference materials are closely connected to the analytical problems arising in the characterization of materials, we will give some topical examples of materials analysis.

a) **Quality Assurance in Production Control.** One of the central tasks of material analysis e.g. in steel and nonferrous metals production, is the analysis of starting and raw materials, like ores, of intermediates and of final products. For the latter the tolerances of certain elemental contents, fixed in standards and norms, must be strictly maintained. In order to control the production process via fast feedback, fast analytical methods like X-ray fluorescence spectrometry (XRF) and optical emission spectrometry (OES) with spark discharge, have become of primary importance.

For example, in the routine production of containers for the transport of radioactive materials, the boron content in iron must lie within narrow tolerances (1.3% > B content > 1%). High boron contents are advantageous for neutron absorption due to the capture cross section of boron, but on the other hand they lower the strength parameters. For the control of products with normed elemental contents (within the tolerance limits), a precise determination of principal and minor constituents is required.

b) **Detection of Value-Determining Elements in Various Matrices.** When analyzing value-determining elements in various matrices, the main emphasis is on the accuracy of analytic values; examples are the determination of the metal contents of ores or the determination of alloy elements in ferro-alloys, like ferrovanadium, ferromanganese, or ferromolybdenum. The umpire assays required (referee analyses) in this connection must be of a high degree of accuracy, the extraction of representative samples being of special importance. The recovery of value-determining elements like Au, Pt or Ag from electronic scrap material or of vanadium from lignite ashes calls for the development of specific analytical procedures with a high degree of accuracy and the development of suitable RMs (see next section).

c) **Detection of Elements Occurring in Recycled Materials.** In conventional metal production, recycled raw materials are becoming more and more common, while the portion of ore and other natural raw materials decreases. A large amount of semifinished products made of lead, steel, and aluminium are produced using recycled starting materials, e.g. lead batteries or aluminium foils. In this process, certain maximum limits must not be exceeded for some elements. The maximum limit of approx. 1 μg/g for Se and Te in Pb accumulators must not be exceeded, otherwise the service-value (*Gebrauchswert*) of Pb cannot be maintained; similar restrictions apply for the maximum limit of the problematic element Pt.

In the production of aluminium foils for the packaging industry, the maximum limit of hazardous elements, like Cd, Zn, Hg etc., must not exceed the concentrations prescribed in standards. Thus the increasing use of recycled materials demands a continuous development of new analytical procedures and RMs, an additional challenge being the specific requirements coming from environmental and health protection.

d) Detection of Trace Amounts in Iron and Steel Industries. As a consequence of new results from materials science and of more restrictive regulations in environmental protection, it is often required to know the contents of Zn, Sb, Pb, B, Ca and other elements in the lower µg/g range in steels; the contents of environmentally hazardous components in slags and powders must be known to even lower orders of magnitude. The requirements of industrial chemical analysis entail the certification of trace amounts in the RMs for iron and steel industries; such RMs are therefore increasingly being introduced. The detection of very low elemental concentrations in certain materials, such as microalloyed steels, has also become important, since low elemental concentrations very often significantly influence properties.

For the control of buildings, bridges or gas containers by means of safeguard inspections, knowledge of the chemical composition of the materials and of their long-term change in time is necessary. In order to perform the relevant analyses in a number of diverse matrices, specific requirements concerning the analytical procedure and the corresponding matrix RM must be met.

e) Elemental Analysis in the Production of High-Purity Materials. Supplying analytical procedures for trace and ultra trace analysis is of importance not only for the semiconductor industries (Si, Ge, AIIIBV, AIIBVI, compounds as well as process chemicals like etching solutions), but also for the production of many other materials. Increasing use is also made of magnetic (superconducting), electrical, mechanical, optical and other material properties, which are essentially determined by the degree of purity.

From the examples given above, one can make the following observations concerning the chemical analysis of materials:

- Classical methods, like gravimetry, titrimetry or coulometry are of decreasing importance for routine materials analyses. Due to their primary character, traceability to SI units and often very high precision, these methods are still very important for the determination of principal and minor constituents in certain analytical problems and specifically for the certification of RMs.
- Routine analyses in production control require ready availability of the analytical results. The goal to be achieved is a direct solid sample analysis by means of optical emission spectrometry with spark discharge (OES) and X-ray fluorescence spectrometry (XRF).
- Modern methods for the analysis of solutions, by means of ICP-OES, ICP-MS or AAS in different variants are used, especially for cases where low detection limits must be reached and a fast feedback of results is not necessary.

4.1.2   Importance of Reference Materials in Materials Analysis

In the preceding section, we gave examples which demonstrate the greatly enhanced importance of materials analysis in an age of globalization of technology and commodity exchange and a fast growing role of high technology. Hence, materials analysis is an essential component of modern technologies.

In order for materials analysis to fulfill this key role, it must be continuously adapted to the diverse requirements coming from all application areas. An essential aspect is the accuracy of results reachable for a specified analytical problem. This accuracy is an expression of the reliability within which the material is characterized and thus of eminent importance for statements concerning its applicability and practical value.

As a rule, modern analytical methods used in the characterization of materials reach such a high precision that it is entirely sufficient for the reliability of the statements. Repeated analyses over shorter or longer periods of time, in conjunction with suitable procedures for correcting the temporal drift of the results, can deliver valid information on precision without using reference materials.

The situation is different when testing the trueness of the results, the second quantity besides precision that determines accuracy. Generally speaking, the analytical methods without primary character, or at least those which must be calibrated, cannot be used without reference materials for problems of this kind. In this regard, reference materials are used for various tasks, but in principle they secure the results by tracing them back, within a specified uncertainty of the results. They promote confidence in analytical results and their mutual acceptance, in particular between the producer or the provider, and the user or the customer. This is possible only under the premises of a mutual acknowledgment of the quality of the CRMs, which must be international and worldwide in the present age of globalisation. RMs enable the comparability of results, or according to De Bievre: "Comparability of chemical results is the goal, traceability is the tool". To this end, the validation of methods and the use of CRMs are necessary instruments.

The use of RMs for various problems is dealt with elsewhere. From the point of view of materials analysis, these problems can be presented in the following way:

- validation of analytical procedures
- calibration of analytical instruments
- testing of the actual accuracy of an analytical procedure
- proficiency test of a laboratory.

These items will now be dealt with in more detail.

Validation of Analytical Procedures

One step in the validation of an analytical procedure is the proof of a sufficient conformity between the results for the analyte contents of a CRM achieved by use of the analytical procedure under investigation and the certified values. In the field of materials analysis one typically has matrix RMs for which analyte content data are not available without analytical measurements (see Sect. 4.3.3). In other words, only in very few cases is it possible to synthesize materials of this kind with well-defined contents. This is mainly due to contaminations and losses, both during the melting processes used for preparing compact samples (metals/alloys, glasses) and in the doping of powdery materials (ceramic powders, raw materials). Therefore the validation of a newly developed analytical procedure can generally only be achieved using an RM, if there already exist other validated procedures for the analytical problem at hand, by means of which this RM could already have been certified. This is often the case, if a new measuring principle is to be introduced, e.g. for faster or less expensive analyses, and if the analytical procedure is not to be extended to new families of substances, other analytes or concentrations or uncertainty ranges. Using this particular CRM in the validation process, one indirectly compares the results with those from other, already validated analytical procedures. At an earlier stage, the quality of these procedures is transferred to the quality of the CRM by means of the interlaboratory certification test; thus one obtains a measure for comparing the quality of the new analytical procedure with the other procedures. On the other hand, such a comparison could also be made by a simultaneous comparison of methods, using an arbitrary material from the relevant substance family having passed a homogeneity test. In the latter case, the comparison of methods (and hence of laboratories) is direct, while in the former case, it is indirect, with the CRM acting as a mediator. Provided a suitable CRM is available, the first mentioned procedure has certain advantages and is easy to handle, since part of the work was already done when the CRM was certified. In addition it allows (indirect) comparison with a large number of different methods that contributed to the certification. Some disadvantages, however, could be that the CRM may not be completely adapted to the analytical problem at hand, e.g. in the sense that not all relevant analytes are certified, or that the analyte concentrations lie outside the range of application of the procedure to be validated. It can also happen that the CRM does not have the required matrix composition. Careful tests and a large amount of experience are needed in order to decide whether and to what degree results obtained with such a CRM are sufficient for the corresponding step in the validation procedure of the method. However, in the field of materials analysis these general analytical problems mainly concern raw materials, additives, etc. or novel and specific high technology products because the concentration ranges of the intermediate or end products to be measured are rather narrow. A further disadvantage is due to the fact that agreement of the results is tested only in one point of the multidimensional concentration space, if only a single CRM is used. It would therefore be desirable to have at least two CRMs with different analyte concentrations for the validation of a procedure. This condition is much easier to fulfill in metallurgy, where one usually has sets of CRMs for one type of alloy, than in the field of high-performance ceramics, for example, where there exist very few CRMs. The role of a CRM in a validation process is to trace the particular method back to other validated methods. In certain cases, error propagation may cause problems. Thus CRMs indirectly serve their fundamental purpose of making analytical results comparable, true and traceable when used for the validation of new analytical procedures.

Calibration of Analytical Instruments

In materials analysis, calibration with CRMs is of special interest for methods that use solid analytical samples without steps of chemical dissolution. The calibration can be done directly and indirectly.

a) **Direct Method.** In this case calibration curves are used directly resulting from usage of CRMs of a certain class of material, e.g. a type of alloy. If the corresponding analytical procedures are routinely used, an uneconomically high consumption of the expensive CRMs may occur.

b) **Indirect Method.** Often a consequence of this expensive consumption of CRMs is the production of internally produced and analyzed calibration samples ("internal calibration samples"); their contents are determined using the calibration curves of the original CRMs. A metrological relation between the internal samples and the CRMs is thus established. Any further analyses can be carried out using the internal calibration samples. This metrological tracing back needs not necessarily be established by means of the method to be validated. Such indirect calibration with CRMs used only as mediators in the calibration will, however, lead to enhanced error propagation and therefore to a greater uncertainty in the result. Sometimes this uncertainty can only be maintained within the admissible limits if the contents of the derived calibration samples are determined with great care and a good deal of experimental input. A further disadvantage with internal calibration samples is the need for an extensive homogeneity test, which for CRMs has already been done. An advantage of the indirect method, besides being less expensive, is sometimes stated as the excellent comparability of internal calibration samples with the samples for analysis produced in the same company, especially regarding matrix composition and further parameters (e.g. grain structure) which influence the results. A continuous improvement of the methods used in materials analysis (see Sect. 4.1.3) has, however, considerably reduced the sensitivity of methods with respect to differences between internal samples and external reference materials. Any falsification of the results due to the use of external samples for the calibration will of course show up in a similar fashion when analyzing internal calibration samples using external CRMs for calibration, unless an additional analytical method is employed, which is independent of the mentioned matrix effects. Any errors in the contents data of the internal calibration samples are then transferred to all results of routine analyses. It can be advantageous to use CRMs from different producers; among other things one can obtain a better estimate of the influence of matrix effects.

c) **Partial Calibration Support by Means of CRMs.** A special form of indirect calibration with CRMs is to first record a calibration curve with internal calibration samples, which have been analyzed by another suitable method previously. The calibration curve is checked in one or only a few concentration points by means of external CRMs. In the case where there are significant deviations, corrective quantities corresponding to a rotation or a shift of the calibration curve are introduced. This is a special case which is particularly relevant if the available CRMs do not sufficiently subdivide the concentration range of one or several analytes. In principle, this special case amounts to the case described before, namely an analysis of internal calibration samples by use of external CRMs; however, for the calibration, the shape of the calibration curve is determined by the results obtained from the internal calibration samples.

The direct or indirect usage of CRMs for calibrating analytical methods in materials analysis transfers the high-quality contents data of the CRM to the calibration curves and thus significantly influences the trueness of all analytical results.

Checking the Actual Accuracy of an Analytical Procedure

The current accuracy of an analytical method may have to be tested by means of a CRM in various situations. In the two cases described below, the CRM can be applied directly or indirectly. As in the previous section, the term "indirect application" will mean that a test sample was used, the content of which was determined by means of a CRM, while the term "direct application" stands for the CRM acting as test sample. The advantages and disadvantages of indirect application are analogous to those described in the previous section.

a) **Introduction of a Validated Method.** In the first case an already validated method (usually a normed method or one described in a collection of regulations) is to be introduced in a materials analysis laboratory. This case is similar to the method validation described previously. The difference is that the validity of this method was already tested before and must be tested in relation to specific laboratory conditions. Any deviation of the results determined in the laboratory from the certified values of the CRM cannot be ascribed to the analytical method in general, but only to the specific application in the particular laboratory. If significant differences appear between certified values of the contents and the data obtained in the practical application of the method, any discrepancy between technical details of the validated method and its practical application must be carefully checked and reduced, until satisfactory results are obtained when determining the contents of the CRM. For this purpose, it is common practice to work with the CRMs themselves.

b) **Continuous Testing of Accuracy.** In the second case, the accuracy of the analytical procedure is tested continuously. For this purpose, a CRM or (frequently in routine procedures) a test sample analyzed by calibrating with CRMs is regularly analyzed after certain periods of time in the same way as analytical samples. In the case of significant deviations with respect to the certified contents, the causes for this behavior must be found and eliminated. As already described in other cases above, use of single CRM or test sample is somewhat problematic. If the calibration curve is rotated with respect to the true calibration curve, it might happen that the concentration of an analyte lies close to the center of rotation so that no change would be noticed. To control such a change it is preferable to use two samples, one in the upper and one in the lower concentration region. These test samples are not to be confused with those used for testing or installing the general performance (e.g. calibration of wavelength or mass, general signal intensity, signal precision etc.). Such instrument-specific test samples are very rarely CRMs. This also holds true for the test samples for temporal drift, which is measured after certain periods of time during a measurement cycle.

CRMs are thus used for the first tests of a newly introduced validated analytical method in a laboratory and for routine tests of the accuracy of established analytical procedures; CRMs therefore play an essential role in the quality assurance of materials analysis, and this is one of the important fields where CRMs are frequently applied.

Proficiency Tests for Laboratories

Concerning their accreditation, the proficiency proof of analytical laboratories being able to carry out certain chemical analyses by means of appropriate tests has become very important, in particular in those fields of analysis which are regulated by law (e.g. clinical or food analyses) and in the whole field of environmental protection analyses [4.2]. In the case of control of effluents, waste water or fly ashes, materials analysis laboratories may also be directly concerned with problems of environmental analytics. But also within the intrinsic tasks of materials analysis, industrial and in part also research laboratories aim at external quality evaluations in addition to the system of internal quality management. For this purpose, CRMs can be used in various ways, as described below.

a) **Proficiency Testing by Special Interlaboratory Comparisons.** The accrediting institution organizes an interlaboratory test and distributes pre-characterized material; the participants must find the analyte contents within prescribed limits. Because of their high costs, one only rarely uses commercially available CRMs, preferring specially produced interlaboratory test samples. Such regular interlaboratory tests, well known from other fields of analysis, have so far been relatively uncommon in materials analysis. One of the reasons is certainly the high diversification of materials analysis and consequently a lower number of potential participants than for interlaboratory tests in the fields of water, waste water and soil analyses. But in connection with certification interlaboratory tests, proficiency tests of the participating laboratories are often carried out, in order to assure their qualification for participating in the final certification interlaboratory test. For this purpose, unlabeled CRMs may be used.

b) Participation in Certification or Validation Interlaboratory Tests. If the laboratory to be evaluated has taken part in interlaboratory tests for the validation of methods or the certification of RMs, the results obtained in this laboratory can be used as a proof of quality; the evaluation limits must be fixed by the accreditor in collaboration with the experienced laboratory. This kind of indirect proficiency test is common in materials analytics. If the results from certification interlaboratory tests are used in this process, the role of the corresponding CRM becomes evident; the importance of CRMs for validation interlaboratory tests was explained above.

c) Bilateral Proficiency Tests. In this case, the laboratory obtains only once a precisely characterized sample, the contents of which must be determined within a prescribed uncertainty. The sample may also be an unlabeled CRM. Due to the generally precise contents data of a CRM, there are distinct advantages over selections of other test materials, when judging the results of the appraised laboratory. This procedure is particularly wide spread in fields where a great number of well characterized CRMs is on the market, e.g. in metallurgy.

d) Comparison Tests in Responsibility of the Laboratory. These tests – also common in the field of materials analyses – are often carried out by using CRMs and comparing the results obtained with the certified values. Thus, the proficiency of the laboratory with regard to a special analytical problem may be assessed as part of its quality assurance system.

Conclusion

The multitude of possible uses of CRMs in materials analysis corresponds to their current importance and wide spread distribution. Many important branches of modern materials analysis, like metallurgy or the glass industry, would be unthinkable without the use of CRMs. But for other special fields and problems, there is an enormous demand for CRMs, which is difficult to satisfy and then only in the long term. Among these materials, there are mostly CRMs in which either low trace amounts must be determined by means of complex methods or for which the contents of rare analytes, barely accessible to analysis, must be certified; or finally materials, which can be reliably analyzed only after an extensive preparation of the sample. For the corresponding analyses, there is often a lack of methods validated with respect to all criteria. An additional economic problem arises, especially if there is a high demand for regulation for some material in the commodity exchange, but if there exist only a few producers who have a relatively low annual consumption of CRMs, in spite of large tonnages. Production and sale of such CRMs are not only unprofitable for potential producers of the CRMs, but often connected with severe financial losses. It will often be difficult for producers of the materials to manufacture these CRMs on their own. It is then up to national or international institutes like NIST and BAM or IRMM (SM & T) to decide whether such economic losses, sometimes pronounced, can still be justified in view of the original tasks of these institutions to support the national economy or regulate the international commodity exchange. If a particular CRM cannot be produced, the commodity exchange of this material must be based on an agreed application of universally accepted analytical methods. Due to a lack of checks by means of a CRM, wrong analytical results may be obtained; law suits and hindrance of the commodity exchange may be the consequence.

4.1.3   Reference Materials for Selected Analytical Methods

As a general rule, an RM should be optimally adapted to the analytical problem for which it is to be used. This means, on the one hand, that the particularities of the applied analytical method are fully taken into account, and on the other hand, that the product should have a high similarity with the samples to be analyzed, if one deals with matrix RMs. Due to the broad variety of samples to be analyzed, fulfilling the second condition will result in a compromise. For the producer of CRMs, this compromise will consist in selecting a RM to be certified which corresponds to a frequent and wide-spread analytical problem in materials analysis, that needs to be regulated. Because of the defined limits of the contents of final products, a broader fluctuation with respect to the sample composition of one type of material is restricted to raw materials, additions, certain starting products (in particular from recycled materials) and the like. Users must select a suitable CRM for their particular analytical problem from the choice of materials on the international market. Their decision will not only be influenced by the composition, form and further properties of the CRM, but also by the general acceptance among users and among their trade partners (for the comparability of the results), and finally also by the price to be paid for the CRM.

Materials analysis has the particularity of being dominated, in many areas of materials production, by fast direct analytical methods. This is because in many processes, especially in melting processes, a frequent and fast feedback is required in the form of the current analytical data. Characteristic methods for this, wide spread in metallurgy, are optical emission spectrometry (OES) with spark discharge (SD) and X-ray fluorescence spectrometry (XRF). By means of these methods, it is possible to analyze compact samples after a quick mechanical sample preparation. Therefore the corresponding CRMs for these methods are compact samples for direct analysis. As in many other fields of analytics, modern analytical instruments for liquid samples have become very common, in particular inductively-coupled plasma optical emission spectrometry (ICP-OES), which has more or less replaced wet chemical methods. Also atomic absorption spectrometry (AAS), already wide spread before the introduction of ICP-OES, is being partially replaced by the latter. Since the samples in materials analysis are usually solids – such as compacts, powders or granulates – they must be subjected to a wet decomposition before being analyzed by methods of liquid analysis. Acid decomposition without pressure is generally used. In order to avoid unnecessary, time-consuming steps, carrying the additional risk of contamination, often mechanically pretreated CRMs reduced to small pieces are available for the methods of liquid-sample analysis, certified in parallel with compact CRMs of the same material. In the following we will mention some peculiarities concerning the relations between methods and CRMs.

The Connection Between Reference Materials and Methods in Materials Analysis: Direct Methods (Solid-Sample Analysis)

Optical Emission Spectrometry with Spark Discharge (SD-OES). Optical emission spectrometry (OES) with spark discharge has developed and changed considerably during the last few decades [4.3]. Middle-voltage spark discharges in argon atmospheres have become the main source of excitation, used in conjunction with polychromators. These polychromators are simultaneous emission spectrometers for a simultaneous analysis of all analytes that are to be determined. They are vacuum instruments or instruments flushed by argon, in order to make the vacuum-UV region accessible for the determination of important nonmetals (like C, P, S, N, and recently also O) and also for further elements having their main wavelengths in this range. The time needed for the analysis itself, i.e. until the results are obtained on a computer, is less than one minute with these instruments; it amounts only to a portion of the time elapsed between sample taking and feedback, which may be several minutes. The times needed for sample taking, sample transport, mechanical treatment of the spark burn area and presparking time are included in this calculation. The spark stands are usually designed for flat samples a few cm in diameter; using a suitable counterelectrode (e.g. a tungsten electrode) a small amount of substance (in the mg or 10 mg range) is removed from these samples in the sparking burning spot during the short time interval, approx. 20 s, which is used for signal registration. Before the analysis, the sample is pretreated mechanically (by lathe, or for harder metals like steel, by grinding). The precise pretreatment procedure is fixed in the prescription for analysis of the various materials. During the presparking phase preceding the sparking phase used for analysis it is now possible by use of modern instruments to perform an "isoformation" of the material. This involves repeated micro-local remelting processes at the frequency of the spark, which for this purpose can be enhanced (high repetition prespark). Higher energy discharges are often used during the presparking phase (high energy prespark) in which a remelted spot is produced that is larger in diameter than the burning spot of the following analytical sparking phase. During this process, differences in the microstructure of different samples are levelled out to a large extend in many cases. These differences result from the different prehistories of the various samples, connected in particular with the kind of casting and solidification of the melt. Generally, it has become possible by means of this isoformation to compare internal samples (made in the company of the analytical laboratory) with external CRMs, without additional time-consuming remelting processes bearing the risk of analyte losses or contaminations.

To sum up CRMs for OES with spark discharge must fulfill the following requirements:

- Since spark discharge is an electrical phenomenon, the materials that can be analyzed are – with very few exceptions – metals and alloys. Because of the geometry needed for spark burning, the CRMs are usually cylindrical, with a plain surface.

- The strong dependence of the intensities of the emission lines from the entire composition of a sample is the main reason why separate sets of CRMs with graduated contents of analytes are necessary for each type of steel or alloy, the application of which must normally be restricted to analysis of that type of material they are produced for. The CRMs are not only used for validation and testing of analytical procedures. Because of the need to calibrate using similar matrix CRMs they are also used for direct calibration or analysis of derived samples with graduated contents used for the calibration.

- By means of OES with spark discharge, reliable analytical results are obtained for a limited concentration range, from medium trace amounts to the region of minor components. Higher contents often cannot be determined accurately enough due to the curvature of the calibration curves. From below, the concentration range is limited by some 10 µg/g (in some cases one order of magnitude higher or lower). General statements are difficult due to the strong dependence of the sensitivities of the main detection lines on the element and due to the variety of possible line interferences. One can conclude that, depending on the kind of material, the CRMs suitable for OES with spark discharge mainly have certified contents between 10 µg/g and 10%.

- The layer thickness of the "effective sample", due to spark ablation, is 10–30 μm, the spark burning spots (a few mm in diameter) can be distributed laterally on the plain surface of the sample. Therefore high homogeneity requirements must be fulfilled by the sample in the lateral (radial) direction. In the axial direction, the consumption of the sample extends over the largest longitudinal range due to the repeated mechanical conditioning of the samples. Consequently the contents of the CRM sample must be constant in this direction as well. During the certification procedure of a RM, this requirement is taken into acount by means of extensive homogeneity tests, usually by OES with spark discharge, and by discarding certain parts of the intermediate products of a CRM, e.g. the ends of a cast rod. With compact samples for spark OES, homogeneity presents a special challenge, because – even when using parallel sparkings – a small sample mass eroded by the spark must be representative of the total sample. Detected radial distributed inhomogeneities may lead to prescriptions concerning "forbidden radial zones" of a CRM (like certain regions in the core or the outer surface ring).
- In spite of the positive effect of structural isoformation during the presparking process, CRMs should not have any segregation (due to solidification or gravitation) or other inhomogeneities, e.g. shrinkholes or blowholes.
- In special cases, it is also possible to use chip CRMs for spark discharge analysis. The user must take suitable measures leading to a compact sample, such as melting or special pressing techniques. Afterwards, these samples are to be pretreated mechanically like other compact samples.

X-Ray Fluorescence Spectrometry (XRF). Besides optical emission spectrometry (OES) with spark discharge, X-ray fluorescence analysis (XRF) [4.4, 4.5] is most common in the area of process analysis of materials. Like OES with spark discharge, this method is mainly used in production control and for the final analytical control of large numbers of solid samples. In stationary instruments, the X-ray fluorescence radiation is usually induced by the primary radiation coming from an X-ray tube. The spectrometers used can be wavelength-dispersive or energy-dispersive instruments. Because of its high spectral resolution and low detection limits, the wavelength dispersive method is the preferred one for many tasks of material analysis. In spite of several technical innovations, the elements with low atomic masses (i.e. the lightest 5–10 elements of the periodic table) either cannot be determined at all or their detection limits are relatively high, when instruments designed for routine analysis are used. For all other elements, the detection limits are as a rule somewhat higher than with OES with spark discharge. But these disadvantages are balanced by certain remarkable advantages, and as a result, OES with spark discharge and XRF complement each other in metal analysis. A great advantage of XRF is the precision of the results, which is determined mainly by counting statistics. If the elemental contents are high, the precision of the results is especially high, and hence random errors may be restricted to a mere 0.1%. Precision analysis can thus be carried out in particular for the main and the minor contents. Increasing the counting times beyond one minute, the range used in routine analyses, precision can be even further enhanced. Similarly to OES with spark discharge, a large portion of the total time between sample taking and the final result is taken up by sample manipulation and sample transport when using XRF. But for both methods there are tendencies towards arrangement of spectrometers nearer to the process and in less central positions, thus shortening the time for sample transport. The preparation of a smooth sample surface with small roughness depth is of utmost importance; this can be achieved by suitable mechanical treatment (treatment at the lathe, mill-cutting and grinding or polishing). Because of the radiation pathway in the spectrometer, samples must be planar. Ideally one has cylindrical disks with diameters up to 50 mm. Smaller diameters are admissible when using radiation apertures. However, this procedure is of limited use because of the smaller area irradiated. The effective information depth in the sample is essentially determined by absorption processes. Thus there is not only a relation with the matrix composition, but also with the analyte and the fluorescence radiation used in the measurement. The information depth is expressed by the "maximal exit depth", at which, depending on the case considered, just a fraction as small as 0.01% of the total fluorescence radiation can leave the sample and contribute to the total intensity. Practically speaking, this depth corresponds to the "infinite layer thickness", i.e. the sample thickness which already gives the maximal pulse rate and a further increase of which would not increase this rate. In the most unfavorable case, the information depth is in the range of a few μm or around 10 μm, which may happen when determining an element with low atomic mass embedded in a heavy matrix, as in the case of the determination of C or Na, respectively, in metals. When determining heavy analytes in a less heavy matrix, this value is significantly higher (up to the range of 100 mm), and even for the detection of Cd in aluminium, the information depth is approx. 5 mm. For different analytes in one and the same matrix, different effective information depths will be obtained.

XRF is suitable not only for metal analyses, but also for analyzing non-conducting substances. In this case, the use of OES with spark discharge is restricted to exceptional cases after pressing powdery samples with electrically conducting powders. Powdery samples can be analyzed directly using XRF. For this purpose they are pressed, in most cases together with bondings, to form flat, solid samples. Using this kind of analysis, considerable grain size effects must sometimes be tolerated. Therefore in material analysis this technique is mainly applied in the process control of raw materials or additives subjected to fast premilling processes. Another application is the routine analysis of waste or secondary products, like slags. Using finer milling will decrease grain size effects; below about 40 μm, there will in general be no intensity decrease caused by mutual shadowing of the particles. The influence of the pressing pressure on the surface, and thus also on the results, is known and must be taken into account during the preparation.

More accurate results can be obtained for powdery samples which were subjected to a digestion by fusion with a suitable fluxing medium, such as lithium tetraborate or lithium metaborate. The samples can be raw materials, additives and waste products, but also intermediates and finished products, like ceramics, glasses, or refractory metals. Using this technique it is not only a disappearance of grain size effects that is obtained. An excess of fluxing medium (up to a factor of 20) is often added, and the sample is in a glassy melt with highly diluted initial matrix. As a result, matrix effects are weakened, and the analytes are distributed homogeneously. On the other hand, generally increased detection limits are observed due to the dilution. This technique can be successfully applied whenever this effect and also the significantly increased time for sample preparation appear to be tolerable. The fusion pellet can be used directly as the sample to be analyzed.

The following requirements may be concluded concerning the properties of CRMs for XRF:

- In contradistinction to OES with spark discharge, using XRF also allows analysis of electrically non-conducting samples. Hence, it is not only metallic CRMs, but also CRMs consisting of ceramics, cement, glasses, raw materials and waste products (ores, slags) which are of interest here.

- As for OES with spark discharge, the analytical signal intensities are highly dependent on sample compositions; therefore for every type of alloy or steel appropriate sets of CRMs with graded analyte contents are also needed for this method. In this case too, CRMs do not only serve for validating and controlling analytical procedures. Since XRF belongs to the class of nondestructive assay techniques, CRMs are directly used for calibration even more frequently than with SD-OES, and they are used to a lesser extent for establishing metrological connections to derived calibrating materials.

- Assuming adequate preparation of the sample, XRF yields accurate results in a concentration region ranging from contents of approx. 50 mg/kg to the concentrations of principal components. For comparison of results, the very high precision is of enormous importance. Due to the widely overlapping concentration ranges, CRMs used for OES with spark discharge in metallurgy can also be used for XRF. But in addition, CRMs are needed for analyzing high contents and the principal components. Particularly for these CRMs, the requirements concerning the resulting uncertainty of the certified values are extremely high.

- In metallurgy (or for glasses), CRM samples are mostly disk-shaped or cylindrical, with a plain and smooth surface, i.e. the metallic samples for OES with spark discharge can normally also be used for XRF.

- The sample volume effectively contributing to the formation of the analytical result of a measurement is small. Especially when analyzing analytes with low atomic weights embedded in a heavier matrix, only the regions very close to the surface will be accessible to the analysis. Therefore these CRMs must fulfill particularly high homogeneity and surface requirements. Unlike in the case of SD-OES, prescriptions regarding "forbidden radial zones" cannot be given, since in general a larger portion of the sample surface is irradiated. Differences in the texture and structure between metallic analytical cast samples and CRMs caused by transforming processes of the starting products for the CRMs may lead to intolerably large analytical errors when using XRF.

- In addition to compact metallic or also glassy samples, powdery samples can be used as CRMs, too; these comprise final products (ceramic powders), raw materials (ores), additives and waste products (slags). Their particle sizes must be sufficiently small (if possible < 40 μm), if they are to be applied in fast direct analyses in the production process. The samples to be analyzed should have a grain size distribution comparable to that of the CRMs used for calibration. This requirement is not relevant when digestions by fusions are applied.

- A speciality of XRF is the possibility of calibration with pure single-component materials, when applying the fusion method. For this purpose, the corresponding pure and, if possible, certified substances are weighed in, and by means of a subsequent fusion, calibration samples with suitable matrix and analyte contents are prepared. The sample to be analyzed is subjected to the same fusion. A method offering particular metrological advantages is the so-called reconstitution analysis [4.6], in which the analysis sample is simulated (usually iteratively) as a fusion pellet with respect to all relevant contents by the calibration sample. Because of the high similarity between the calibration and analysis samples, any disturbing matrix effects are largely compensated. The pure substances needed for the calibration should be available as CRMs or should be metrologically bound to CRMs. If they are used for simulating the matrix, the trace concentrations of all relevant analytes are of importance. If in turn they are used to calibrate the analyte, the principal component must be certified with extreme accuracy. Thus these pure-substance CRMs must fulfill stringent requirements concerning an accurate and as far as possible comprehensive characterization by elemental analysis (see Sect. 4.3.1).

Glow Discharge. Laser Ablation. Both techniques are used for the ablation of samples and are coupled to atomic spectrometric detection, with special emphasis on aspects of locally resolved analysis.

Glow discharge OES (GD-OES) can be applied in a similar way to OES with spark discharge for bulk analysis in metallurgy. Due to the low-pressure character of the discharge, larger linear dynamic ranges can be achieved, in particular towards higher concentrations. Thus for this method also CRMs can be used which would be more suitable for XRF because of their contents in minor components. Owing to the low sputtering rates, the homogeneity requirements are similar to those for OES with spark discharge. In spite of a number of advantages, GD-OES did not become anything like as common a bulk method as OES with spark discharge. Due to its approximately plane layer ablation of the sample, GD-OES is used advantageously for depth profile analysis, in particular for layer analysis. The signal dependence on sample composition and ablation depth must be calibrated, but using CRMs without a layer structure, this cannot always be done with satisfactory accuracy. Problems arise from the sputtering process and from limitations of available quantification methods [4.64]. Certified multilayer stacks of known layer composition and thickness can be used for calibration, optimisation and determination of depth resolution, as well as for checking the accuracy of the quantification method. A modern development is high frequency GD-OES making possible the analysis of electrically insulating materials. However, there is a significant lack of suitable CRMs with a layer structure. An ISO study group on the availability of thin film standards appropriate for GD-OES depth profiling was established in 1996. A list of criteria for a reference material to be useful has been worked out [4.65] and a report on the availability of suitable reference materials based on these established criteria will be published [4.66].

In 1997 a new VAMAS project on GD-OES depth profiling was initiated by the Bundesanstalt für Materialforschung und -prüfung (BAM), Berlin, Germany. The overall goal within the VAMAS project (Versailles Project on Advanced Materials and Standards) is the development and definition of the design of depth profiling standards (Ti/Al-multilayers onto 100Cr6 steel and SiO_2/TiO_2-multilayers onto BK7 glass) for GD-OES. Procedures for intensity calibration and the determination of depth resolution will be worked out and as a result the production and the certification of such reference materials will be organized by BAM. This is of great significance especially in the field of modern high-tech materials. There is a significant lack of suitable CRMs with a layer structure.

Coupling glow discharge with MS (GD-MS) results in a direct detection method with extremely low limits of determination mainly for electrically conducting samples, but one which is very complex regarding instruments and which involves time-consuming operation thus rendering it less common. The sample ablation is significantly less than with GD-OES. CRMs must therefore fulfill significantly more stringent homogeneity requirements with respect to the sample depth. The main application area is the extreme elemental trace analysis, if time-consuming steps carrying the additional risk of contamination are to be avoided. For this particular task, very few CRMs are available. Concerning the trueness of the method, the situation can be improved by also using CRMs with contents above the extreme trace region and by extrapolating the calibration curves.

The coupling of laser ablation with ICP-OES or ICP MS (see below) results in methods for laterally locally resolved microanalysis (approx. 10–300 μm); the instruments are available optionally together with ICP spectrometers. They are suitable for the direct analysis of compact, electrically conducting and nonconducting materials and are rarely used in a routine manner in material analysis. The CRMs for this method must have a composition similar to the samples to be analyzed. Essential micro-fluctuations of analyte contents of CRMs over longitudinal distances in the order of magnitude of the crater diameters must be excluded. Since no special CRMs exist, the CRM material used must be tested for extremely high homogeneity in each case. When analyzing glasses or glassy fusions, for which these problems do not exist, special attention must be paid in order to have a similar absorption coefficient for the radiation used, both for the CRM and the analysis sample. With appropriate scanning, the methods are also suitable for statements on the entire bulk contents of the sample or for extended regions. In this case, there are no extreme requirements regarding microhomogeneity.

Analogous statements hold for other methods used for local analysis, which have in part significantly higher local resolution, like electron or ion microanalysis or SIMS, for which no specific reference materials exist either.

Special Methods for the Determination of Non-Metals. When analyzing metallic samples, methods of atomic spectrometry, in particular OES with spark discharge, can also be used for determining some non-metallic analytes. Important examples are the determination of C, P and S, but also of Si and As, in steels. Increasingly, nitrogen and oxygen contents are also determined by OES with spark discharge. In several cases, however, the limited detection power of the methods of atomic spectrometry for the light, technically relevant elements H, C, N, O and S will be a problem. Furthermore, the applicability of these methods strongly depends on the availability of suitable CRMs. Thus it may be advantageous in these cases to use the widespread methods of combustion analysis and of carrier-gas hot extraction. Recently CRMs have been developed for oxygen determination in copper materials for hot extraction/combustion analysis in the lower concentration range on the one hand, and for OES with spark discharge in the higher concentration range on the other hand. This clearly reflects the tendency towards an increased use of OES with spark discharge, in addition to the classical methods.

A number of different techniques are used in combustion analysis/carrier-gas extraction for determining C, H, N, O, and S. Depending on the technique and instrument type, weighed sample portions of 1–200 mg or of 10 mg to 1 g are recommended. The dynamical range of the methods is high and extends from the mg/kg region to the 10 g/kg region. The contents range covered by different CRMs is correspondingly large. In addition to solid samples, solution residues and gases can also be analyzed. For solids, this fact often makes a primary calibration possible, based on solutions or admixed gases and independent of CRMs. In these cases the use of matrix CRMs is often reserved for testing the trueness and for the calibration in case of routine analyses.

In certain cases, powdery materials can be problematic as CRMs because their reactive surface areas can be quite large. A typical example is the change of the oxygen content in metal and metal carbide powders during storage [4.7]. Chip-shaped and similarly shaped CRMs are in principle suitable, as well as rods or wires, from which appropriate volumes can be cut. In most cases, their surfaces must be subjected to a substance-specific chemical cleaning procedure (etching). During this process, the formation of porous or rough surfaces is to be avoided. Massive metallic CRMs should also undergo a suitable mechanical cleaning before a sample portion is removed. Otherwise a part of the oxide skin can get into the interior of the sample during the treatment, as has been observed when treating soft metals at a lathe. If activation analytic methods (in particular photon activation analysis) are used in conjunction with hot extraction/combustion analysis, as is done sporadically in institutions with the corresponding equipment, there is an essential advantage in the sense that after activation even strong cleaning procedures can be used leading to larger and more reactive surface areas. The reoxidation of the sample is done by means of non-activated oxygen, which is not detected. This may be an important advantage especially for the certification of reference materials.

Feedback times may be short when the conventional analytical methods are automated, and this is used e.g. for steel production control. Therefore the contents of C, Si, P, and S in compact, powdery or chip samples of steel CRMs for multi element analysis are certified up to very few exceptions, and often data are also given for the N contents. In addition there are special steel CRMs for the exclusive determination of O and N.

Methods with Foregoing Wet Digestion. Normally the methods for analyzing liquid samples require a time-consuming sample preparation and cannot often be used as fast methods for process control in materials analysis or for a quick final control of large piece numbers. But they have nevertheless retained their traditional importance in materials analysis. The range of methods chiefly used has shifted tremendously in recent years. Classical methods of primary character, often metrologically superior, were superseded by faster modern methods, usually by those of atomic spectrometry. Due to their nonprimary, comparative character, these methods depend to a much higher degree on suitable calibration substances and CRMs.

General Aspects. The methods of liquid sample analytics differ from those of solid sample analytics in essential points concerning the way and the necessity of using CRMs. This is mainly due to the fact that solutions with complex composition and defined concentrations can easily be prepared and homogenized, simply by mixing separate solutions. The problems arising in the preparation of homogeneous massive or powdery samples are thus avoided.

Assuming mutual chemical compatibility and sufficient solubility, solutions of almost arbitrary composition can be prepared. From this also follows, that by a simple mixture or dilution of monoelement solutions and chemicals used for digestion, one can simulate matrix concentrations as they are found in a digested sample. By adding defined portions of monoelemental solutions, one can prepare a series of calibration solutions of defined analyte contents with graded concentrations. The necessary degree of matrix matching depends on the analytical problem, on the required accuracy of the result, and on the method applied. Analogously to preparing solutions for this external calibration, one can also increase the concentrations of analytic samples with defined analyte contents by spiking according to standard addition procedures. In this case the problems resulting from incomplete matrix matching are nonexistent. One can conclude, that in most cases, independent of the particular kind of calibration, a calibration can be achieved by simple means which is sufficiently adapted to the problem by using monoelemental solutions. Since the calibration solutions can be prepared gravimetrically, starting from pure substances of defined stoichiometry, or ideally starting from materials as elements, the calibration can be traced back directly to the SI unit. In this case, the true purity of the substance must be known. (For further comments concerning this set of problems see Sects. 4.3.1 and 4.3.2.) If the samples are digested by means of fusion for use with XRF as described above, advantages exist similar to those just described. Glassy samples have properties analogous to liquid samples, even though formed by a fusion process, which may be quite complicated.

Besides the advantages for the calibration in solution analytics, some limitations also apply:

- If, after the digestion of the analysis sample, the solution contains organic components influencing the result of the analysis, these components must also be taken into account. In materials analysis, however, such problems arise much less frequently than in environmental analysis, for example; they may play a role nevertheless in organic matrices, like plastic materials.

- The solutions resulting from complicated digestion procedures may contain undefined amounts of digesting agents or matrix compounds with these agents, which cannot be simulated by means of simple reactions occurring by mixing of solutions and digesting agents. If the results are disturbed by this, the matrix solution required for the calibration must be generated through digestion of a relatively pure sample of similar matrix composition, containing the relevant analytes at irrelevant low levels. Such materials are, however, in many cases not available. Alternatively one may use an analytical procedure with low detection limits and analyze highly diluted samples. This is a general procedure for reducing certain matrix effects (the nonspectral ones in atomic spectrometry). Problems may arise due to blank values falling into the range of the low concentrations obtained by dilution.

- In materials analysis in particular, direct solution analytic procedures without matrix separation are often used. In such solutions, the matrix has the same excess concentration with respect to the analyte as in the solid sample. Applying external calibration, the required matrix matching often presupposes the existence of very pure substances. These may often be unavailable, and in addition the contents of the relevant analytes contained in them must be determined for each individual case (see Sect. 4.3.1).

In spite of these restrictions, analytical methods using liquid samples offer advantages for calibration. The disadvantages are not only enhanced time consumption, but also the existence of additional blank values, contaminations and losses of the analytes during preparation and handling of the samples.

We now want to consider the question of the purpose of matrix CRMs in materials analysis, since calibration is generally achieved directly using mixed pure solutions that are traceable to pure substances or pure CRMs. Concerning this question the following statements can be made:

- The availability of pure CRMs of defined stoichiometry, if possible elemental, in the solid state and with well certified main element concentration is of great importance for the traceability of the analyte calibration using these methods. However, there is a significant lack of sufficiently certified materials of this kind. This is a great disadvantage, particularly in precision analysis (see Sect. 4.3.1 for further remarks on these problems).
- The situation is similar for analytics in the trace region, concerning the existence of ultrapure CRMs with certified trace amounts for matrix matching.
- In both cases metrologically derived solutions can also be used. Two points deserve special attention when using these solutions: first, the acid contents of the calibration samples must not differ from those of the analysis samples; second, the stated uncertainty for the content of the principal component or the stated concentrations of the trace amounts (in case of matrix matching), respectively, must result from a metrologically correct traceability chain (cf. Sect. 4.3.2).
- In analytics with solutions, the calibration with solutions of ultrapure substances constitutes a second metrological cornerstone in addition to the use of matrix CRMs; with the exception of XRF of fused digested samples or of NAA with ultrapure solid substance calibration, there is nothing like this in the analysis of solid substances. This second metrological cornerstone is not only fundamental for the final determination of the sample, but its influence also affects the sample preparation to some

extent. This is true in the sense that analyte concentrations added in a controlled manner to the analysis sample before the sample preparation, either as solutions or as weighed-in ultrapure substances, must be recovered as a correspondingly increased final value, if there are no losses during the sample preparation. But since the chemical and physical states of the analytes in the "natural" and in the spiked part of the sample may be different, these recovery experiments are only of a limited value.

- This limitation is the main reason for the continuing and irreplaceable importance of the matrix CRMs for solution analytics, at least in cases, when the sample preparation steps are rather complicated. Using matrix CRMs, the trueness of the total analytical procedure can be checked within the precision of the method and within the uncertainties of the certified values of the CRM. As in other cases, a sufficient similarity is necessary regarding the contents of the analysis sample and the CRM. This in turn presupposes the existence and availability of sufficiently many CRMs for the needs of materials analysis.

- These matrix CRMs for materials analysis should be available in such a form that they can be dissolved without any complicated mechanical sample preparation carrying the additional risk of contamination. For this purpose, chip-shaped, pellet-like, shot/grit-shaped or powdery samples have proved useful. The chips must be small enough and well enough mixed to yield a representative partial sample, when taking samples commonly in the range of 0.1 g to 1 g.

- In some cases powdery samples may have disadvantages, because, if the individual grains differ too much with respect to composition or grain size, demixing may occur. In material analytics this may happen with CRMs of raw materials and of waste products. A further disadvantage with powdery samples is the altered stoichiometry of regions close to the surface, due to chemical reactions during storage. In most cases this affects only the determination of oxygen or other nonmetallic analytes, which as a rule are not determined by methods of solution analytics. A falsification of the mean bulk stoichiometry by surface changes, which could lead to noticeable analytic errors through a change of the reference mass, is to be expected only in the case of particularly reactive materials with a large specific surface area. A preconditioning of powdery CRMs by heating is to be recommended in many cases, however, in order to remove adsorbed surface humidity.

In the following passage some details and peculiarities are discussed concerning the most widespread analytical methods for liquid samples in the field of material analytics.

Optical Emission Spectrometry with Inductively Coupled Plasma (ICP-OES). The ICP-OES is a very widespread method with multi-element properties [4.8–10] and is applied increasingly in the field of material analytics. More than 70 analytes – mainly metals or semimetals, but also S and P – can be determined sequentially or simultaneously and sometimes even in both modes depending on the type of spectrometer. The sample throughput using a simultaneous spectrometer with about 30 samples per hour is approximately 3 times higher than for sequential instruments. Recently it has become possible to reduce considerably the sample consumption of 3–30 mL obtained using common nebulizers to 1 mL or even less with the help of special microconcentric nebulizers. In this way, amounts of CRMs may be greatly reduced when used for purposes of analytical quality control. Using a matrix salt concentration of 10 g/L and relating to limits of determination in the solutions of about 10 μg/L the lower limits of the applicability of the ICP-OES for many analytes related to the solid samples are calculated to about 1 μg/kg. With this detection power most tasks of modern material analytics with regard to trace analysis of metallic analytes are fulfilled. Because of the high linear dynamic range of up to 5 orders of magnitude, concentrations at least up to 1000 mg/L are determinable. The dynamic range can be further extended to higher concentrations by dilution of the liquid samples or by using less sensitive spectral lines. Therefore trace and minor constituents of material samples can be determined at the same time. The errors caused by limited trueness and precision lie mostly in the range 1–5%. Hence the method is less well suited for precision analytics of high contents. But precisions distinctly below 1% can be achieved by taking appropriate measures like further sample dilution, use of an internal standard element, repeated measurements with drift control and application of a modern precision nebulizer. The trueness is influenced to a much greater extent by spectral interference than by non-spectral interference [4.11]. Thus, even if the analytical spectral lines are carefully selected and spectrometers with high resolution (5 pm) are used, the possibilities of analytical application of ICP-OES concerning the determination of several analytes are limited, when handling matrices rich in spectral lines (e.g. Fe, Cr, Ni).

Atomic Absorption Spectrometry (AAS). The AAS [4.12] is a mono- or oligoelement method and yet remains widespread in the material sector, not least because it is relatively economical and well established with many validated analytical procedures. For the 70 or so determinable analytes, limits of determination at a level of a few mg/kg are typically calculated in relation to the solid sample material when the flame technique is used, compared with those of about 0.1–1 mg/kg when the graphite furnace technique [4.13] is used. The lower sample throughput of AAS is one of the main reasons that has led to a partial superseding of this method by ICP-OES, especially in cases when larger numbers of analytes are to be determined. But compared with ICP-OES the high selectivity of AAS is an advantage. Besides this, the high precision of flame-AAS (about 1%) and the higher detection power of the graphite furnace AAS should be emphasized. The possibility of hydride generation AAS is also of importance in the field of material analytics for the determination of elements appropriate for a hydride forming reaction, in which an in situ separation of these analytes from the matrix takes place. If possible, AAS should used for the certification of a material as a CRM in addition to ICP-OES. Because these methods are based on different analytical principles, experience shows that systematic errors in the final determination are often only discernible if both methods are used.

Mass Spectrometry with Inductively Coupled Plasma (ICP MS). ICP MS [4.14,15], which can be used to determine more than 60 elements in one fast sequential mode, has become more widespread in the field of material analytics during the last decade. The linear dynamic range is still higher than with ICP-OES, but the matrix tolerance is lower (approx. 1 g/L maximum). Nevertheless low limits of determination can be achieved for solid samples of 10 μg/kg and even lower, because of the extremely low limits of determination for solutions at a typical level of 10 ng/L. Apart from the quadrupole ICP mass spectrometers double focussing sector field mass spectrometers are becoming in use. Using these instruments at their highest resolution of about 10000, many sources of interference related to the presence of molecular ions, which render difficult or even impossible the determination of certain isotopes using low resolution, can be overcome. In material analytics, there is also a growing need for determining contents at the lower mg/kg level and below. This need results from new knowledge concerning the relations between performance characteristics of materials and low contents of certain additive elements. This is relevant for common materials, such as steel, as well as newly developed materials such as high performance ceramics. Another trend comes from the need to acquire analytical information on an ever larger number of trace elements. Apart from new knowledge in material engineering, this also arises in the context of recycling processes, where analytes not customary in raw materials are introduced into the products. In addition new regulations and resulting demands from the enviromental or life sciences concerning the prohibition of potentially toxic trace elements are also contributing to this trend. Considering these developments, ICP MS will certainly become more common in the material sector since it is mainly used in the field of trace analysis. Some urgently needed CRMs with certified contents at the lower trace level are now under development by recertification of existing CRMs, e.g. for the steel industry, or by certification of new RMs for hightech materials, such as high performance ceramics.

4.2   Preparation

The history of RMs in the iron and steel industry goes back to the early part of this century when the first cast iron RMs were prepared in 1906 by the National Bureau of Standards (NBS) in the USA in conjunction with the American Foundrymen's Association.

Six years later the Königliche Materialprüfungsamt in Berlin, the precursor institution of the BAM of today, issued a "normal steel" made of non-segregated material for the carbon determination. The program was extended rapidly so that already in 1913 eight samples with different certified carbon contents were available, and in 1914 samples with different certified Mn contents were added, ranging from a simple manganese steel via cast iron (*spiegeleisen*) to ferromanganese. By 1920, steel RMs with certified S, Cr, Ni and W contents had been added. Then 21 different CRMs were offered by the precursor institution of BAM. The production of iron, steel and nonferrous RMs was enhanced by BAM at the beginning of the 1950s.

With the encouragement of the NBS the British Chemical Standards (BCS) movement was started in 1916 by Ridsdale & Company to prepare RMs (or standard samples as they were then called) of iron and steel in cooperation with analytical chemists in the industry. Later this work was expanded to include non-ferrous alloys, raw materials and refractories, and in 1935 it was taken over by Bureau of Analysed Samples Limited (BAS) under the directorship of N.D. Ridsdale, one of the founders of the organization.

In 1922 the Société Française des Echantillons-Type was founded in France and it prepared iron and steel RMs up to the early 1940s. In 1953 the work was resumed in France by the Institut de Recherches de la Sidérurgie Française (IRSID).

Production and certification of RMs for metallic materials and their starting materials on a high metrological level is mostly a task for state institutions (e.g. BAM, LGC, NIST, NMi), industrial associations (e.g. JCS, VDEh, GDMB, CTFI) or private institutes. There are also international communities producing CRMs, like the European Union (IRMM) or the EURONORM group for the production of RMs in the iron and steel sector, consisting of IRSID (F), CTIF (F), BAS (UK), BAM (G), VDEh (G) and the MPI für Eisenforschung (D). An expert group of the European Committee for Iron and Steel Standardization (ECISS) acts as consultants to the group of producers. A list of abbreviations is given at the front of the book.

4.2.1 General Remarks

CRMs of a high metrological level and reliability are usually prepared by independent state institutions or with their assistance (ISO/REMCO 420) [4.16]. Since in general the starting materials are not produced in this institution, but are obtained from producers of special products (ISO/REMCO 425) [4.17], the confidence in the analytical result is enhanced by separating the producer of the candidate material from the independent certifier (CRM producer). The reader can find more information about production and use of RMs in [4.18] and [4.19].

Sometimes it is necessary for the certifier to modify the candidate material, e.g. by means of tempering or doping. For homogenizing the material, special pretreatments are required, depending on the material properties, like machining (sometimes at low temperatures), grinding, sieving etc. In certain cases and in close cooperation with the producer, special alloys of homogeneous metal samples are prepared by fast cooling, thus avoiding segregation processes.

The multi-element methods, optical emission spectrometry with spark discharge (SD-OES) and X-ray fluorescence analysis (XRF) require special RMs when used to monitor the production of the metal-melting and manufacturing industry. Being relative methods, they must be calibrated against samples of precisely known contents, which must be as similar as possible to the material to be analyzed. The homogeneity required for RMs can be achieved by solidification of the material from the melt for many metallic alloys, but due to segregation, inhomogeneities cannot be sufficiently avoided for certain alloys (cf. also Sect. 4.2.3). Pressure-sintered steel samples of sufficient homogeneity and reproducibility, for example, could be produced by means of a powder metallurgical procedure, namely hot isostatic pressing of powders pulverized and solidified under inert gas.

That this method works was shown, for example, with the heat-resistant steel X85CrMoV18 2 and the high-speed steel S 6-5-2-5 [4.20,21]. Neither material differs in its signal curve from the common steels, solidified as bulk material from the melt, when investigated by means of SD-OES and XRF; they therefore are suitable for producing CRMs for both methods.

The method explained above can in principle also be transferred to other systems; it thus becomes possible to produce homogeneous alloys for RMs which are otherwise difficult to obtain.

On the other hand, special RMs are also manufactured by renowned producers of certain product lines. Examples are metal RMs of the steel and nonferrous metal industries (pure metals, alloys). The metrological level of these products may well be somewhat lower unless they are certified by interlaboratory tests according to internationally recognized guidelines. The chemical composition is usually only determined by means of one or several methods in the manufacturing company itself (examples come from aluminium industry). Another problem is the fact that the certifying institution will not generally have a neutral attitude towards evaluation of the material.

For further information concerning reference materials like definitions, quality, uses etc. see the appropriate ISO guides 30–35 [4.22–27].

4.2.2 Principles of Reference Materials Preparation

CRMs can be prepared in several different ways. According to Rasberry [4.22–28], there are three possible ways for producing certified reference materials, among which NIST prefers the second possibility (ii). The three possible ways consist in using:

(i) a previously validated reference method;
(ii) two or more independent, reliable measurement methods;

(iii) a network of cooperating laboratories, technically competent and thoroughly familiar with the material.

BCR (BCR guidelines A and B) [4.29,30] and BAM (BAM guidelines) [4.31] prefer the third possibility for certifying RMs.

The development and production of RMs are closely related to industrial product and material development. As a consequence of the analytical tasks connected with this development, there is a continuous demand for new RMs, and this demand must be determined periodically via inquiries. Once a producer has decided to prepare an RM, following (iii) he has to proceed according to ISO Guide 35, BCR and BAM guidelines for the preparation of CRMs and take the steps described below for the preparation of an RM by means of interlaboratory comparison. The route normally taken (item (iii) above) also corresponds to an IUPAC recommendation. A list of producers of reference materials can be found in (ISO REMCO 330) [4.32].

4.2.3   Preparation of Certified Reference Materials According to BCR and BAM Guidelines

When producing a CRM, the following steps must be taken, as summarized in the scheme below:

- feasibility study,
- selection of candidate materials,
- homogeneity and stability tests,
- interlaboratory comparison,
- certification procedure, statistical assessment,
- certification report,
- certificate.

We will now give more details about the individual steps leading to the preparation of an RM, with special emphasis on the aspects of materials analysis.

Feasibility Study

A feasibility study is first of all aimed at a demand analysis for a future CRM. This study must contain the necessity and scope, and also the group of prospective clients. From this study one can estimate the number of CRMs to be produced. There must also be an estimate of the risks connected with the development of a CRM, like financial expenditure and time consumption. The risks also comprise the availability of candidate materials and their costs, preliminary data on stability and homogeneity, and identification of laboratories willing to participate in interlaboratory comparisons. Producers must also examine whether they have the capacity for manufacturing a large amount of

sample material, e.g. for the subprocesses of grinding, sieving, homogenization, and sample partitioning. In the case of a positive decision, further steps
will follow. In materials analysis, this decision is often taken by national or
international bodies. Examples are the GDMB and the EURONORM-CRM
group (see Sect. 4.2.). Another possibility is for a competent material producer, well acquainted with the market and products, to prepare the CRM
at a relatively low risk.

Selection of Materials and Pretreatment

From a number of candidate materials, the producer of the CRM must select
those which are to be included in the certification procedure via an interlaboratory test. The candidate materials are usually obtained from a single
producer (e.g. from the steel, aluminium or chemical industries) or, in some
cases, are produced in the company itself, which is natural, if material and
CRM producer are identical.

Already during the selection process, the homogeneity requirements of an
RM must be taken into account. If one uses metal alloys for example, prone to
segregation, special measures for enhancing the homogeneity of the compact
material must be taken, e.g. by means of extremely high cooling velocities or
by forging and milling the original ingots down to lower diameters. Dense and
homogeneous material samples can also be obtained by hot isostatic sintering
of finely divided powders.

When using powders or chips, special attention must be paid to use only
one batch which will be portioned via a sample divider. Following a special scheme for mixing subportions by means of a rotation sample divider
(cross-riffling), the homogeneity of powder and chip samples can be enhanced
considerably. For methods like XRF and OES with spark discharge, the manufacture of compact samples is the goal, while for wet chemical procedures,
powders and chips are most suitable. Thus any unnecessary preparatory steps
by the RM user will be avoided. Often a large number of candidate materials
must be tested, of which only a few will survive the extensive preliminary
tests, because they fulfill the requirements for a future RM. The necessary
analyses are performed by the producer with selected batches, sometimes
together with other laboratories, and sufficient reserve material will be retained.

Homogeneity and Stability

The amount of sample used for a single measurement is essentially decisive for
the homogeneity requirements, which the material has to fulfill. If metallic
RMs are produced from rod-like material, the homogeneity tests are carried out by means of horizontal and longitudinal single measurements. Disk-
shaped material is particularly well suited for the OES and XRF methods;

when using OES, the homogeneity requirements are particularly stringent, since only a few mg of sample material are evaporated during spark burning.

When used for elemental determination from solution, the material samples, which have already been tested for homogeneity, must be chipped after removing the casting crust or mill scale. In most cases, the chipping is done by turning and milling [4.33]. In pretests, the optimal cutting parameters are determined for obtaining sufficiently fine and uniform chips. The cutting velocity must be selected in such a way that an excessive evolution of heat is avoided during dry machining, i.e. without auxiliary cutting tools. The chips must be as finely divided as possible; 1 g of the sample must consist of 400 to 1000 chips. Thus minor inhomogeneities still existing in the starting material can be compensated after thoroughly mixing the chips.

The mixing of the chips in done in drum mixers. Fine portions whirled up during the mixing are sucked off. The drum mixing will also round sharp edges on the chips. Thus a later crumbling off of fine particles is largely avoided. After drum mixing, the chipped material is introduced into a sieving machine for separating the desired chip size from other sizes.

When preparing chip samples, a thorough removal of the fine-grained portions and precautions against their later generation are very important measures to be taken. In many metallic structures, it is the microcrystalline phases which tend to break off, and their composition deviates markedly from the average composition. These fine-grained portions would gradually accumulate at the bottom of the container and thus affect the homogeneity of the sample.

The risk of sedimentation is particularly pronounced with powder materials, e.g. ferrous alloys, ores, slags and other powders. In the ground material, the grain-size distribution is determined, and then each grain fraction is separately analyzed chemically. For the final RM which is to be certified, only grain fractions with a very low sedimentation risk are selected. To be on the safe side, users of such RMs are notified that before each sample taking the whole material must be thoroughly mixed.

In spite of all precautions, materials like alloys prone to segregation or gray cast iron, where graphite deposits easily break off the metal structure, cannot be transformed into sufficiently homogeneous chip samples. In such cases, spraying the raw material from the melt will be useful. Each solidifying droplet has then the same composition as the melt. Due to its high cooling velocity, cast iron will solidify white, i.e. without any graphite deposits. Sometimes, however, even in sprayed material the composition depends on the droplet size. This may result from surface-active components of the melt. Therefore it is mandatory to carry out an extensive sieve analysis, to analyze the individual grain fractions separately and then to select the fitting grain-size classes for the future RM.

The spray method not only assures a very homogeneous starting material, but also has the advantage of avoiding the considerable time expenditure required for chipping the material (100 kg of chips correspond to 500 machine hours). If the melt contains volatile components, e.g. cerium, the method is only of limited use, since time-dependent evaporation losses can occur in the melt during the relatively long spray procedure.

When producing inorganic nonmetallic powder RMs, homogeneity is of primary importance, and the materials must be ground, sieved and homogenized by repeated sample divisions.

The actual homogeneity tests consist of two steps (within-units homogeneity, between-units homogeneity), where the variances belonging to analytical results from different sample ranges are compared statistically.

For testing the homogeneity, an analytical method with a high precision should be used. The trueness of the analytical method is less important due to the relative data evaluation. The precision of the analytical method should be high enough to ensure that relative inhomogeneities can be recognized as such.

1st Step: Homogeneity Within Unit of a Batch. Here the variance within the unit, i.e. within a single flask with powders or chips or within a single compact sample is determined. The average variance within a representative number of units is compared with the ideal variance which would result when analyzing a perfectly homogeneous sample. The error appearing in the analysis of such an ideal sample would only be determined by the precision of the analytical method used. If the true variance is significantly larger than the ideal variance, there will be an inhomogeneity within the units.

2nd Step: Homogeneity Between Units of a Batch. Here the variance between the mean values of individual selected units (converted to single values) is compared to the average variance within the unit. If the variance between the units is significantly larger, this homogeneity test yields an inhomogeneity between the units.

When an inhomogeneity is found, the following steps must be taken:

- if the inhomogeneity is very large, the samples cannot be used as CRMs;
- a significantly determined tolerable inhomogeneity can be included in the total uncertainty budget;
- in certain cases the material can be certified individually.

Although long-term stability plays a minor role for inorganic materials as compared to organic materials, the stability period as stated in the certificate must nevertheless be assured. In particular, it must be ascertained whether demixtion processes in metals or interactions of powders with the ambient air (reaction with CO_2, H_2O or O_2) could lead to irreversible changes and weight changes.

Proficiency Test

If the previously described preparatory steps have been completed so that the selected prospective candidate material can be certified by means of an interlaboratory test, a proficiency test will be needed to prove the competence and well documented operating procedure of the laboratories participating in the interlaboratory test. For this purpose, samples together with given reference values for the different elemental contents are distributed among the selected laboratories. At retrieval rates within set tolerance intervals, the laboratories have proved to be competent and may thus take part in the certification test. If the materials are difficult to analyze or are hard to decompose, the proficiency test can be repeated or, if necessary, carried out with increasing complexity.

Interlaboratory Comparison

The samples tested for homogeneity and stability will be sent by the responsible institution to the interlaboratory test participants who have passed the proficiency test. It is important that as many methods as possible, based on different measuring principles, should be applied. The criteria of trueness, precision, repeatability, reproducibility, comparability and traceability must be fulfilled by each independent analytical method.

Assessment

The statistical evaluation of the analytical results obtained by interlaboratory test participants via different methods is the task of the coordinator of the interlaboratory comparison. Under certain conditions, the data sets obtained by one laboratory with different methods can be treated as data sets of different laboratories. The data sets of an interlaboratory test can be evaluated by means of various methods.

According to the BCR guidelines [4.29], "pooling of all individual data" (i.e. all sets of data produced by the various laboratories may be considered as samples from a single population of data and therefore treated as one single data set) is recommended only if the means and variances do not differ significantly.

In the case of "no pooling", the BCR guidelines use a very simplified model for the calculation of the certified value. This means that laboratory mean values are modeled and not individual values.

In Sect. 4.4.6 we give an example for the evaluation of an interlaboratory test including the application of outlier tests for the preparation of a porous CRM.

When evaluating the data sets of individual laboratories, outlier tests must be carried out and the reasons for the observed deviations must be clarified (e.g. incorrect instrument calibration). Important outlier tests are for example the following [4.34]:

- Kolmogorov–Smirnov–Lillefors test, which tests the coincidence of the distribution of the individual results and the laboratory mean values with respect to the normal distribution.
- Cochran test, by which outliers in the variance of the individual laboratory data sets are determined.
- Dixon test, which tests whether an individual measured value is an outlier value under the assumption of normal distribution.

Statistically outlying results should be subjected to a careful examination of all aspects but should not be rejected merely because they are outliers. The fact that a set of data is in disagreement with the others does mean that it is wrong in most, but not in all cases. Before deciding to reject an outlier, the participants should have carefully investigated the possibility that the outlier is the only correct value.

The result of the evaluation is to be stated in the uncertainty budget, which contains, among other things, standard deviations and confidence regions.

Certification Report

The certification report contains all details concerning preparation and application of the CRM, in particular data belonging to the uncertainty budget, such as:

- mean value of laboratory values,
- standard deviation of laboratory mean values,
- standard deviation between laboratories,
- mean standard deviation within laboratories.

Furthermore, the report contains data concerning the analytical methods used by the participant laboratories, etc.

If there is no agreement of the results after excluding outliers and if the reasons for this disagreement are unknown or if the variety of methods is insufficient, certification must be excluded.

If the agreement of the results obtained with different analytical methods is insufficient and if the reasons for this discrepancy cannot be elucidated, a certification must again be excluded.

Certificate

The producer of the CRM prepares a draft of the certificate and discusses the results with the participants of the interlaboratory test. If all participant laboratories approve of the draft, the certificate is passed. In addition to the certified values, and the tables with the laboratory mean values, it contains the names of the participant laboratories with anonymous assignment of the values and the applied methods.

In special cases, also non-certified guide values and further information will be given for individual values of the chemical composition or other properties (e.g. phase composition, particle-size distribution, etc.).

4.3 Special Qualities and Properties (Types of Reference Materials)

In the following passages several mainly metrological aspects are discussed concerning the classification of CRMs in the field of materials analytics.

4.3.1 Pure Reference Materials for Calibration

As described in Sect. 4.1.3, high purity substances or mixtures or solutions made from them in the laboratory can be used for calibrating certain analytical methods. This holds for all methods for the analysis of liquid samples as well as for XRF when using fusion as the digestion technique and also for neutron activation analysis (NAA). By this kind of calibration, traceability to the SI unit is made possible. This explains the exceptional status of high purity CRMs with regard to metrological aspects and why they are often classified using the ambiguous term "primary CRMs". According to the rules of many renowned producers of CRMs, primary methods or methods using defined pure substances (or solutions traceable to them) for the calibration are the only ones allowed for certifying new RMs. This is because calibration with matrix CRMs could result in a confused chain of traceability. In materials analytics non primary methods needing a calibration are ordinarily used. The uncertainty of the contents of the high purity CRMs is of special importance, because this uncertainty is transmitted in the same proportion to the final results where small tolerances are often stipulated. Many validated and normed analytical procedures contain a passage concerning calibration which starts by the dissolution of appropriate pure substances with a defined stoichiometry. To avoid errors coming from a poorly defined stoichiometry, e.g. the aqueous component in metal salts, one will try to use substances in elemental form, normally high purity metals, if possible. When the element is highly reactive, as for alkali metals, one will fall back upon stoichiometrically well defined compounds. Unfortunately, there hardly exist any pure substances certified with respect to the main component with an unambiguous statement of uncertainty. For pure compounds the statements of the producer are mostly limited to some trace concentrations and to a minimum content of the main compound. Often this will be given as "not below 99%". In this example one would have to expect about $99.5\% \pm 0.5\%$ for the most probable content of the pure compound in the entire substance. By taking into consideration also stoichiometric and other uncertainties, the uncertainty of the content of the metallic component may grow to about 1%. In the field of materials analysis such a value is often not tolerated, being the contribution to the entire uncertainty of the result and only coming from the uncertainty of the starting calibration substance. At least, this holds for the determination of minor and of main constituents. Considering the usage of pure metals instead of compounds the situation is sometimes not much more favorable. Concerning the entire trace "metal" content extremely pure substances are offered, e.g. marked by the producer with the quality "6N" (which really means "m6N") corresponding to a maximum sum of all "metallic" impurities of 0.0001%. But the content of the main component is often strongly influenced by the content of non-metallic traces, especially by the oxygen content. If the material has a large specific surface area, as for powders, then intolerably large uncertainties in the content of the main component may arise in spite of a high purity of the material. Only few metallic materials are provided by the producer with stated values related to the total content of all trace impurities. The few existing certified values are mostly in a region of "t3N" or worse, i.e. they are related to maximum total contents of all impurities of 0.1% or higher. For most analytical problems the corresponding uncertainty in the main component will be sufficient, but not in the field of precision analytics. With only few exceptions these substances are offered as "pure substances" and not as "pure substance CRMs" and one has to assume that the stringent rules for certification such as those in the BCR or BAM guidelines, had not been applied in most cases. Because of this the uncertainty in those values could be further increased.

For the purpose of matrix matching another kind of pure substance with appropriate low contents of all analytes is often additionally needed. For this purpose in some cases the CRMs with the lowest contents can be taken from the sets of pure metal CRMs consisting of metallic samples with graduated concentrations of technically relevant analytes, as is usual, e.g. for pure Cu or pure Al. In other cases there is no choice but to fall back on the pure metals not directly declared as CRMs, as mentioned above. The disadvantage lies in the fact that there is often no information given on concentrations of relevant analyte traces or that the values only have an orientating character. An advantage is that the real purity is often sufficient regarding matrix matching. The reason for this is the very high purity concerning the "metallic" analytes, and only these are of importance for the calibration of methods where matrix matching is usually used.

In any case a clear lack of comprehensively certified pure substances can be inferred from the facts explained above. This is true for pure substances accurately certified for the content of the main component (the pure metal) for calibrating the analyte as well as for ultrapurity substances certified for a large variety of relevant trace contents for matrix matching in the field of trace and ultratrace analysis. Efforts have thus been made by leading institutions to remove this lack, concerning the metrological basis of inorganic analysis, of essential importance in the field of materials analytics. Examples of this development are the pure substances certified by the working group "primary substances for calibration" at the VDEh (Verein Deutscher Eisenhüttenleute – German Iron and Steel Institute). The chairmanship of the working group is held by a scientist of BAM. These substances are especially appropriate for calibrating XRF when using the fusion technique for sample digestion. Until now, pure substances of SiO_2, Al_2O_3, $CaCO_3$, Ni, NiO and MgO have been certified with regard to the main component and a greater number of trace contents. While these materials are designated for customary, commercial distribution, other kinds of ultrapure substances are involved in the certification process of the national institutions, which are designated to remain there as the metrological base and for making comparisons and balance with the other institutions. Efforts made in Germany by BAM to develop an internal "system of primary calibrating substances" in cooperation with PTB can be understood in this context. This system will contain two well selected and certified substances, one for analyte calibration and the other for matrix matching, for each nongaseous and nonradioactive element of the periodic table. The main components are certified with an uncertainty of $< 0.01\%$ (partly $< 0.001\%$). The certification of these materials for analyte calibration is carried out by determining the contents of all possible traces (including non-metals and gases); the content of the main component is then the difference between 100% and the sum of all trace contents. The trace contents of all "metallic" analytes of the other type of samples for use as matrix matching materials are certified with limits of determination of 0.001 mg/kg up to about 0.1 mg/kg. The very low uncertainty in the contents of the main components of the substances determined for analyte calibration means that they are also appropriate for use as backspikes in the primary method of isotope dilution mass spectrometry (IDMS). Another effort is being made by the Swiss organisation EMPA, where primary calibration substances of this kind are subjected to a second purification procedure by distillation under high vacuum conditions. The European IRMM intends to integrate these developments into its project of metrological improvement of IDMS.

4.3.2 Standard Reference Solutions for Calibration

In the standard analytical instructions of the leading institutions it has long been preferred to put solid pure calibrating substances at the disposal of each laboratory and to use these as the basis for calibration in every case, thus shortening the metrological chain and avoiding errors due to the limited shelf life of solutions and the limited possibilities of freely adjusting acid concentrations when using ready-made element solutions for calibration. Various disadvantages stand opposed to these advantages in comparison to the use of commercial calibrating solutions which are commonly provided by many producers:

- the difficulty or often the impossibility of making appropriate and certified materials available (see Sect. 4.3.1),

- the greater demands on personnel, finances and time when preparing stock solutions,
- the risk of errors when preparing or diluting solutions gravimetrically or volumetrically,
- the risk of introducing additional contaminations.

These reasons have contributed to the fact that the importance and application of commercial calibrating solutions (with concentrations mostly of 1 or 10 g/L) has grown markedly. The question of the metrological quality of these solutions is of decisive importance because it greatly influences the quality of the results.

There are various possibilities for increasing the acceptance of this kind of solution:

- Checking concentrations by comparison with solutions prepared in the laboratory, which are based on definite high purity substances. This presupposes the existence of sufficiently precise analytical procedures for carrying out the comparison.
- Application of reliable calibrating solutions from renowned producers. For this purpose the calibrating solutions of NIST are widespread and widely accepted, mostly having nominal concentrations of 10 g/L and concentration uncertainties from 0.3% up to 0.7%.
- Comparison of the concentration of calibrating solutions prepared in the analytical laboratory or of commercial solutions without definite statements of uncertainty with those of reliably certified element solutions from renowned producers. The comparisons have to be carried out in the application laboratory which must have at its disposal precise analytical procedures for the metrological comparison. The uncertainties caused by these procedures contribute to the total uncertainty of the results.
- Comparison of the concentrations of element solutions coming from at least two different producers. The concentrations are accepted when they correspond to a sufficient degree. But a defective statement for the concentrations could remain unrecognized if the commercial solutions stem from the same starting material and if no further control is carried out.

Several commercial producers have made efforts towards reliable and sufficiently small uncertainty budgets for their element calibrating solutions. To this end, different approaches are followed. One trend is discernible, however, namely that producers also offer solutions of a higher metrological quality with a reliable uncertainty budget in addition to the customary solutions of the past. The considerably higher prices sometimes required for the new kind of solutions are justified by the fact that time-consuming primary methods are used for their certification, like titrimetry or gravimetry, and that a comparison is made with generally accepted calibrating solutions of national institutes (NIST) to establish metrological traceability. Additional costs also arise by using ultrapure starting substances and by certifying the trace concentrations. Thus these solutions are universally applicable for analytical multielement methods of high detecting power such as ICP-OES and especially ICP-MS for matrix matching. Furthermore there exist models by which national institutes delegate the production of solutions to commercial producers, whilst control of traceability to their CRMs is done themselves (NIST), or the concentrations of solutions produced by commercial partners are certified with defined uncertainties by national institutions (EMPA, BAM) using metrologically high-quality analytical methods.

Besides producing especially well certified and pure calibrating solutions commercial producers are also beginning to offer multielement solutions with concentrations adapted to special methods or even to the specific requirements of one customer.

From the many efforts mentioned above one can conclude that systems of metrological traceability of calibration using element solutions is a very important problem in continuous development and not yet solved by any unique approach.

4.3.3 Matrix Reference Materials

Examples are given in Sect. 4.4 regarding important applications of the so-called "matrix CRMs" widespread in the field of materials analytics. These are CRMs which are similar to the samples to be analyzed with regard to matrix composition and other sample properties. As in other fields of analytics, these matrix CRMs are quite different to those substances or solutions described in Sects. 4.3.1 and 4.3.2 because of their adaptation to one real analytical problem. In Sect. 4.1.3 the strong correlation was pointed out between the analytical methods used, the properties of the investigated material samples and the properties of suitable CRMs. When a generalizing sample preparation step is used for preparation (wet digestion or digestion by fusion) the similarities between analytical sample and CRM sample may be mainly important regarding the sample digestion itself, because the calibration of the final determination can generally be done using solutions (or fused samples) coming from pure substances. But even when using this kind of sample for calibration, the application of matrix CRMs well adapted to the analytical task can be of great importance if the trueness of the result has to be checked (see Sect. 4.1.3). Inevitably certain differences generally exist between an analytical sample and a carefully selected CRM in view of their composition. Further parameters, such as the structure or texture of samples, barely influence the final determination when analytical procedures are used which include a digestion for sample pretreatment. In every individual case one must carefully assess to what extent differences in the concentrations of the main and minor constituents of the analyzed sample and CRMs are contributing to potentially differing results. Depending on the extent of such differences, discrepant results for determined analyte concentrations may also arise when the CRMs are used for calibration in direct analytical procedures. Moreover, the possibility of giving statements regarding the accuracy of an analytical procedure is impaired and reduced. The reason is that the uncertainty in analyte concentrations of CRMs that were determined by those procedures is only transferred to a limited extent to the accuracy expected for analysis of real samples. A typical example for this is the analysis of wet digested samples using a synthetic external calibration procedure with matrix matching. In the ideal case, concentrations of major and minor constituents in the calibration solutions are well adjusted to those of the real sample solutions to be analyzed. If then the contents of a CRM used for controlling the accuracy of the results are found to deviate from the certified ones, this deviation may be much higher than that caused by the uncertainty of the analytical procedure alone, because the calibration solutions are not really adjusted to the concentration of the dissolved CRM but only to concentrations of the dissolved real sample. The concentrations of major and minor constituents in the dissolved CRM may differ markedly from those of the dissolved analytical samples and the adapted calibration solutions, leading to matrix effects. In this case an analytical procedure could be rejected even though it may be of satisfactory accuracy. On the other hand an analytical procedure not satisfying accuracy requirements could be accepted, if the deviations of the determined results for the CRMs caused by insufficient matrix adjustment are accidentally directed in the opposite direction to those deviations caused by really existing systematic errors inherent in the analytical procedure. The latter would in this way be more or less compensated and hidden. From these considerations the necessity of having CRMs very similar to real samples with respect to major and minor contents can be concluded. But one of the peculiarities of materials analysis consists in the fact that the final products normally have lower and upper concentration limits which are rather narrow. Within these limits, sets of CRMs with graduated concentrations are in most cases available. Thus the problem described, of such great importance in other fields like environmental analytics, can be classified as a second degree problem, at least in connection with CRMs for final and intermediate products. The situation is quite different with regard to application of CRMs for controlling analytical procedures used on starting or waste products, which normally have distinctly wider ranges of variations in their contents.

Finally some peculiarities of matrix CRMs in the field of materials analyses are summarized here:

- the very stringent demands to be met by the homogeneity of CRMs applied with direct analytical methods as a consequence of the very low effective amount of sample consumed when using these methods (SD-OES, XRF, hot extraction/combustion analysis);
- the difficulties in preparation of alloying samples by melting processes with sufficiently high lateral and radial homogeneity;

- the influence of secondary parameters (such as texture) on the results, which are often not negligible, and the resulting limited possibilities for using certain CRMs when applying direct analytical methods;
- the advantageously long-lasting temporal stability of compact sample CRMs (striking exception: determination of oxygen).

4.4 Applications

The following chapter gives a few examples of applications of RMs in different industries. The specific problems concerning chemical composition (types of RMs), sample form (powder, compact samples) and application of RMs for different analytical methods are discussed. For different matrices the situation is not uniform. A great number of metallic reference materials (ferrous and non-ferrous metals) are available. For these materials, the RMs are classified in groups of material types of steels and alloys. The relative frequency of those groups of material types is given in several figures, which will be interpreted. Information is given for finding desired matrix RMs in databases (e.g. COMAR database) or distributors of RMs.

For the description of reference materials in other material classes, where only a few RMs exist, another approach has been chosen. For ceramic and glass reference materials, specific examples are given together with the certified element contents.

A specific situation exists regarding RMs for the determination of particle size, specific surface area or distribution of pore sizes. RMs with these specific properties are mentioned here because of their importance in chemical and physico chemical analysis resulting from the relation between the particle and pore size distribution of a solid and its properties. Examples related to these properties are catalysis, separation and reactivity. RMs for the calibration of instruments can only be used in the method specified by the certification. Their use for other methods will give incorrect data. The application of method-specific RMs will therefore be discussed in detail.

4.4.1 Reference Materials for Iron and Steel Analysis

Reference materials are very important for laboratories in the iron and steel industry for various reasons, as described in Sect. 4.1.2 in a wider context [4.35]. On the one hand this applies especially for the confirmation of all analytical results, i.e. for the control of accuracy. On the other hand the determination of reference materials is important in demonstrating the accuracy of analytical determinations to accreditation bodies, suppliers and customers. Even small discrepancies between measured and true values can have serious financial consequences. In Germany, for example, 40 million tons of iron ore are consumed each year. The value of this iron ore depends on the iron content alone. A difference of only 0.1% iron corresponds to the sum of 2 million DM [4.36].

Table 4.2. Certified reference materials, special alloys

CRM No.	Material	Main components	Form of sample	Producer
SRM 1247 (1984)	Incoloy (825)	Ni (43.5%), Cr (20.1%), Fe (46.2%)	Disc	NIST
D 326-1 (1972)	High N high Cr special alloy	Cr (16.37%), Ni (61.16%)	Chips	ECRM producers group
F 378,1 (1995)	Special alloy	Co (63.52%), Cr (28.22%), W (4.43%)	Chips, disc	ECRM producers group
22x 9011 (1998)	Nimonic 901 alloy	Cr (12.65%), Mo (6.10%), Ni (45.1%), Ti (2.10%)	Disc	MBH Analytical Ltd.
2528-98A (1989)	Ni-base alloy	Cr (17.43%), Mo (2.88%), Ti (2.11%), Fe (7.63%), W (3.09%)	Chips	ISO TSNIICHM, Russia
JK 37	Cr-Ni-steel	Cr (26.72%), Ni (30.82%), Mo (3.55%)	Chips, discs	Jernkontoret, Sweden
BCS 398	Almico HC	Ni (16.59%), Al (9.98%), Co (14.92%), Cu (6.09%)	Chips	British Analysed Samples Ltd.
GBW 01501		Ni (80.07%), Mo (4.13%)	Chips	NRCCRM, China

Types of Material

An important part of all reference materials available are reference materials for the analysis of iron and steel products. Approximately 20% of the 10 000 or so reference materials listed in the COMAR database belong to the field of iron and steel products. These materials can be divided into the following types of material:

1. Pure iron. In this type of material the element contents C, Si, Mn, P, S, N as well as Al, Cr, Ni and Cu are of special interest.
2. Unalloyed, alloyed and highly alloyed steels. The division between unalloyed and alloyed steels depends on given limits of contents of different elements and is described e.g. in standard DIN EN 10020 [4.38]. In these materials the elements mentioned under (1) as well as Mo, Co, V, W and Ti are of interest.
3. Special alloys, for example, based on Cr and Ni. In this small group of materials the main elements are those mentioned under (2) and in addition iron which is present in the materials only in the range of a few percent. The iron content is thus often given as a certified value.
4. Cast and raw iron (carbon content > 2%). The elements mostly certified in this group are C (2–5%), Si, Mn, P, S, Cr, Ni, As, Cu, Ti, V and N.

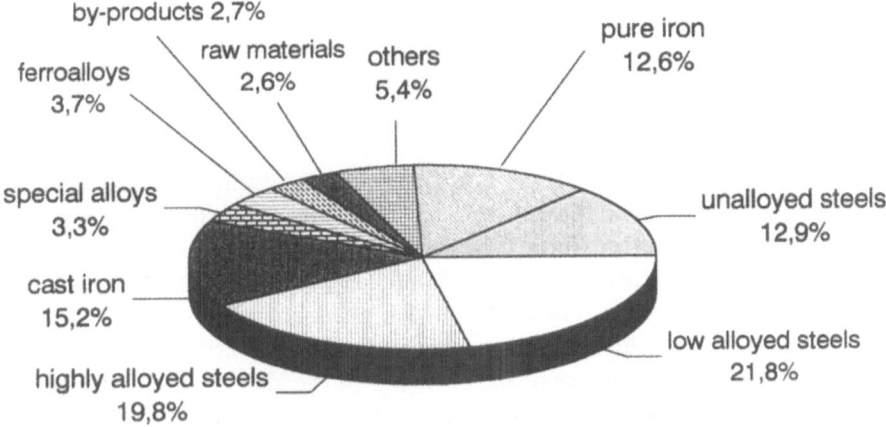

Fig. 4.1. Relative frequency of the different types of reference materials used for steel industry analyses. Data come from the COMAR database and are based on about 1850 entries

The carbon content is much higher in cast and raw irons than in the materials mentioned above (Table 4.2).

5. Ferroalloys. Depending on the type of ferroalloy the main elements, such as Mo in FeMo, Mn in FeMn, Cr in FeCr, W in FeW, etc., which are in most cases the certified elements, are relevant for the price of the material and are the most important. In addition, impurities like C (up to 7%), Si, Mn, P, Cr, Al, Ti and V are of interest.

6. Ores and concentrates. In this as well as in the following group the elements Fe, Si, Ca, Al, Ti, Mg, Mn, P, S, Na, K and Cr are the most important elements.

7. Ceramic materials and minerals (e.g. lime, dolomite), refractory materials (e.g. SiC, magnesite).

8. Slags, dusts, coal. The main interesting elements are those mentioned under (6) and additionally F and U.

In some of the materials given under (7) and (8) the contents of compounds (e.g. TiO_2, SiO_2, Al_2O_3, etc.) as well as additional properties like loss of ignition or water content are certified. Figure 4.1 shows the distribution of the different material types as parts of all reference materials for iron and steel analysis taken from the COMAR database.

Producers of Reference Materials

The most important producers of reference materials for the analysis of iron and steel products today are:

- the National Institute of Standards and Technology NIST (USA);

- the European EURONORM-CRM producers group, consisting of the British Bureau of Analysed Samples (BAS), the French Centre Technique des Industries de la Fonderie (CTIF) and Institute de Recherche de la Sidérurgie (IRSID), the German working group "Certified Reference Materials for Iron and Steel", consisting of the Federal Institute of Materials Research and Testing (BAM), Verein Deutscher Eisenhüttenleute VDEh and the Max Planck Institute (MPI) for Iron Research as well as the Swedish/Finnish Nordic Certified Reference Material Working Group (NCRMWG). The Institute for Reference Materials and Measurements (IRMM, in former times BCR), which produces reference materials for many different fields does not produce reference materials for the analysis of iron and steel because of the existence of the European EURONORM-CRM producers group (exception: coal).
- the Chinese National Research Center for Certified Reference Materials
- the Japanese Iron and Steel Federation (JISF)
- ISO TSNIICHM (Russia)

There are in addition a number of other companies producing reference materials for the analysis of iron and steel or selling their inhouse reference materials. Beside reference materials with well established certified element contents there are samples for recalibration and drift control, for example, for spark emission spectrometers. These standardization or setup samples must be very homogeneous, whilst exact knowledge of the contents of the elements is not so important. When measuring these samples, only the resulting signal intensities are compared with older measurements to exclude any drift effects. There are a lot of producers specialised in production of these standardization samples, such as MBH Analytical Ltd. or SPEKTRARAT and others.

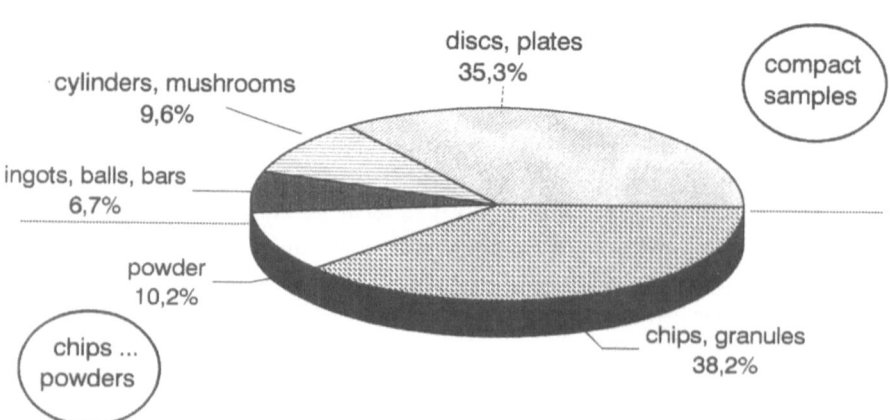

Fig. 4.2. Relative frequency of the different forms of reference materials used for steel industry analyses. Data come from the COMAR database and are based on about 1850 entries

Form of Samples

General remarks concerning the relations between the different forms of CRM and the methods they are used for are given in Sect. 4.1.3. There are in general three kinds of forms used as reference materials for iron and steel analysis (Fig. 4.2).

Compact Samples in Disc Form. Compact samples are used for calibration, recalibration and quality control measurements using spark emission (SD-OES) and X-ray fluorescence spectrometry (XRF). Because staying time of the melt in blast furnaces is still decreasing today there is often not enough time to analyse the melt with classical or wet chemical analysis methods to control the composition of the melt. Therefore samples are taken from the molten metal and these samples are analysed by spark emission spectrometry directly after solidification and cooling of the sample. Another reason for the increasing importance of compact samples is the possibility of determining nonmetallic elements like nitrogen – as a result of technical progress in the analytical instrumentation – with SD-OES. Taking into accout these technical possibilities, there is an increasing number of reference materials in compact form where the producers also certify the nitrogen content. (There exists a programme of the European EURONORM-CRM producers group for certifying the nitrogen content in seven existing EURONORM reference materials.)

When producing compact reference materials it is extremely important to control the homogeneity of the samples. Where samples are produced by casting the molten material into rods there is the possibility of inhomogeneities within the rod, i.e. there are differences between the top and the bottom of the rod from which the samples are produced by cutting the rod into pieces. This can be the case when the casting of the melt takes a long time allowing losses of volatile elements. On the other hand there can be inhomogeneities from the outer part of the sample to the center because of segregation effects during the solidification of the melt. One way of obtaining very homogeneous compact samples is to atomize the melt in an inert gas atmosphere (e.g. argon), followed by hot isostatic pressing of the powder [4.21].

Samples in Chip or Powder Form. Reference materials in chip or powder form are mainly used for wet chemical analysis. They are taken as substances for calibration, e.g for analysis of solutions using atomic absorption spectrometry (AAS) or ICP- optical emission spectrometry (ICP-OES). An important advantage of calibration with reference materials versus calibration with commercial or homemade standard solutions is the fact that any effect coming from the matrix will be taken into account if the matrix of the reference material is similar or equal to the matrix of the sample.

A second important application of reference materials in chip or powder form is – as for compact samples – to control accuracy of analytical methods or control drift effects when measuring large series of samples. Furthermore, reference materials are used for the validation of new analytical methods. On the one hand it must be checked whether the analytical method to be validated is suitable on different types of matrix, while on the other hand important data like concentration range, recovery values, standard deviation of the method and interlaboratory standard deviation must be determined with the help of reference materials. The uncertainty of the reference material must nonetheless be taken into account.

Using reference materials in the form of chips, it is important not to remain below a minimum quantity of sample material (minimum number of particles) because any possible inhomogeneity between different particles will then be eliminated. The problem of inhomogeneities between single particles has already to be taken into account during the production process. For instance it is known that there are inhomogeneities for carbon between particles depending on the particle size of ferroalloys. During the crushing process of carbon-rich cast iron the graphite sometimes deposits. It is therefore necessary to eliminate graphite deposits before conditioning the material as a reference material. In general it must be ensured that there are no inhomogeneities between different particle size fractions, otherwise some of the fractions must be discarded.

If reference materials are used for calibration, it must be remembered that the uncertainty in the measurement increases because the uncertainty in the reference material becomes part of the total uncertainty. Because of the higher uncertainty, calibration with reference materials should be avoided in cases where high accuracy is necessary, e.g the determination of main components in ferroalloys, where the price of the material depends on the content of the main component.

Samples in a Special Form Like Balls, Rods, etc. Determinations of the non-metals nitrogen, oxygen, sulfur and carbon in iron and steel laboratories are normally carried out using analytical instruments based on infrared or thermal conductivity detection after combustion or carrier gas heat extraction of the analyte in a crucible. Nowadays for some of these elements – as mentioned above – spark emission spectrometry is used for the determination. The instruments based on combustion or carrier gas heat extraction are normally calibrated by reference materials. Calibration with pure chemicals or with gas has the disadvantage of not taking into account any matrix effects. Especially for the determination of oxygen, several reference materials are available for calibration of the analytical instruments. The oxygen content in the steel melt has to be checked because this element causes an uncalmed solidification, leading to small bubbles in the solidified steel. The steel is then susceptible to unyieldingness.

Samples in the form of small rods or balls are well suited as reference materials for calibration, because it is easily possible to remove the oxide layer on the surface by chemical pickling before the determination. There

Table 4.3. Certified reference materials for the determination of nitrogen and oxygen

CRM No.	Material	Content [%] oxygen	nitrogen	Form of sample	Producer
ON 201S-1 (1994)	Low alloyed steel	0.0049	0.0429	Balls	Ovako Steel, Sweden
D 029-1 (1970)	Unalloyed steel	0.0312	0.0083	Rods	ECRM producers group
D 099-1 (1987)	Ball bearing steel	0.0008	0.0078	Gold plated balls	ECRM producers group
SRM 1754 (1989)	Low alloyed steel	0.0024	0.0081	Rods	NIST
12 M XAAS	Steel	0.0037	0.024	Discs	MBH Analytical Ltd.
CM 3032	Steel	0.0187	0.0360	Balls	CMSI, China
GS 1C	Steel	0.0046	0.0254	Rods	JSS, Japan

are also a few materials which are plated with a very thin layer of gold (e.g. ECRM 099-1). These materials can be put directly into the crucible of the oxygen analyser without any pretreatment because the surface is inert against oxidation. Some certified reference materials for the determination of oxygen and nitrogen are listed in Table 4.3.

Future Developments

An important tendency in the iron and steel industry is that the manufacturing time, e.g. the time the melt remains in the blust furnace, decreases. Therefore analytical determinations have to be very fast. The result is that instrumental analytical methods are replacing classical analytical methods. Another reason is that the number of employees in laboratories is decreasing, so it is often impossible to use time-consuming analytical methods for element determinations. Many instrumental methods require compact reference materials for calibration, and the significance of samples in chip form decreases compared with compact samples. The disappearance of analytical capacities in industrial laboratories is also a big problem for reference material producers, who establish certified values by round robin experiments and where only wet chemical methods or analytical techniques which can be calibrated with pure substances are allowed for determinations.

In addition to the elements mentioned above other elements are becoming more and more interesting. Examples are:

- Ca in alloyed steels. Added in small quantities Calcium improves material properties with regard to casting and turning [4.38].

- environmentally sensitive elements because of government directives, e.g. for the packing material industry. As these elements are often contained only as traces, the demands on analytical methods used for certification analysis concerning detection limit and the like are increasing. New and sensitive analytical methods like ICP-MS are used for the determination of these trace elements.

In addition to total element contents, interest in other properties is also increasing. An example is the isotopic distribution of boron in steels used in the nuclear industry. Another point of interest is sometimes the species of an element, e.g. the acid soluble and insoluble part of aluminium in steel or the content of Fe(II) in addition to the total iron content in ores.

4.4.2 Reference Materials for the Analysis of Non-Ferrous Metals

The above-mentioned statements regarding the frequency of different kinds of RMs in the field of analysis for iron and steel (Sect. 4.4.1) – as well as those that follow regarding the field of non-ferrous metals – are based on information from the COMAR database. It should be remembered here on the one hand that the database does not include all RMs on the market and on the other hand that, besides the real producers of RMs, different distributors of RMs are also included in the COMAR database. A further problem consists in the fact that it is not possible to search the COMAR database for particular types of alloys. As the basis for the following statements, access was made via the ranges of alloying element concentrations – but this is awkward and involved and not uniquely defined. Another problem is, independently of the quality of the COMAR database, the high diversification of the different

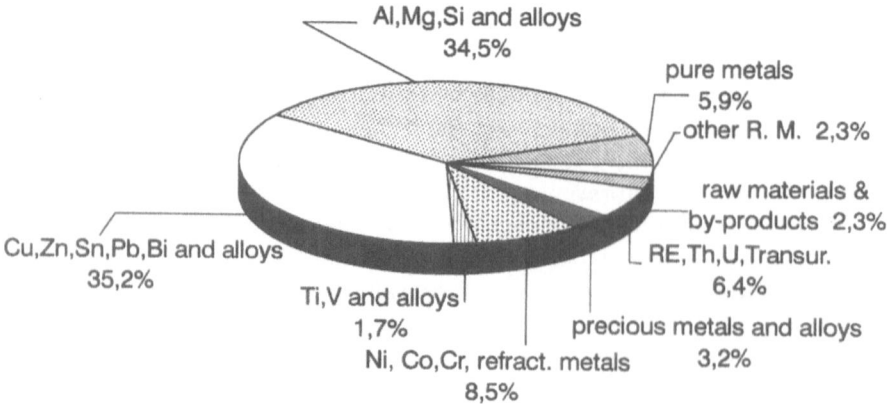

Fig. 4.3. Relative frequency of reference materials used for analyses of non-ferrous materials. The data come from the COMAR database and are based on about 3400 entries

types of alloys render a comprehensive view rather difficult. Nevertheless, regardless of these restrictions, it is possible to give several general statements concerning the main fields of RMs for the analysis of non-ferrous alloys.

Further to the 1850 or so entries in the COMAR database for the analysis of iron and steel products considered in this paper (Sect. 4.4.1, Figs. 4.1 and 4.2) about 3400 RMs for non-ferrous metals are registered in the database. Figure 4.3 shows the frequency of different types of material among these samples. Owing to their technical importance, RMs in the group of Al, Mg and Si (mainly alloys) and a second group of samples based on Cu, Zn, Sn, Pb and Bi (mainly alloys, too) dominate the distribution of all the RMs. More than one third of the total number of all registered RMs for this field are contained in each of these groups. These two groups are described in more detail under the following two subheadings, while the remaining approximately 30% of CRMs are described under the third subheading.

Figure 4.4 shows all RMs encompassed by the COMAR database for the analysis of non-ferrous metals, arranged according to the form of the material. Dominating, with more than one third of the samples, are plates or discs closely followed by cylinders, mushrooms and other spectrographic standards with a little less than one third. The latter category of forms is especially suitable for use with spark emission spectrometry and X-ray fluorescence spectrometry (see Sect. 4.1.3). Except for a disproportionately small remainder, these materials belong to both main material groups of non-ferrous RMs, namely the aluminium and the copper alloys (and the materials assigned to these groups). In the category of plates and discs, which may be well suited for X-ray fluorescence, most samples are also made from these two groups of materials. Approximately 96% of all Al materials and about 80% of the copper materials belong to the two categories of material forms mentioned. This reflects the high degree of application of fast instrumental methods in the corresponding branches of industry.

The production of compact reference materials is made by continuous casting or by extrusion of ingots with a diameter of up to 30 cm or by atomization (spraying through a nozzle) of the melt in an inert gas atmosphere (e.g. argon) followed by hot isostatic pressing of the resulting powder to yield especially homogeneous samples. Reference materials in the form of mushrooms, especially for Al alloys are produced by casting the material in little frames. Other forms like bars or ingots are also produced by this method. By rapid cooling, radial concentration gradients may be avoided.

The remaining third of all non-ferrous reference materials are in the form of powders (9%) for wet chemical analysis (classical methods or ICP-OES/AAS), or in the form of chips and granules (10.6%) also suitable for wet chemical analysis and partly for the determination of gases and non-metalic impurities using heat extraction and combustion methods. For the latter analytical methods ingots, rough globules, bars and wires (5.2%) are especially well suited.

Aluminium and Aluminium Alloys

As already mentioned above aluminium materials are one of the two largest groups of non ferrous reference materials. With Fig. 4.5 aims to give a classification of the frequency of important alloy types based on reference materials with aluminium contents of more than 35%. Because of the large number of different types of alloys only a rough summary can be given. This leads to the following conclusions:

- Most of the certified reference materials, representing more than one quarter of this group, are aluminium–silicon alloys. Besides the contents of the minor constituent silicon with mostly a few % (partly below 1% or up

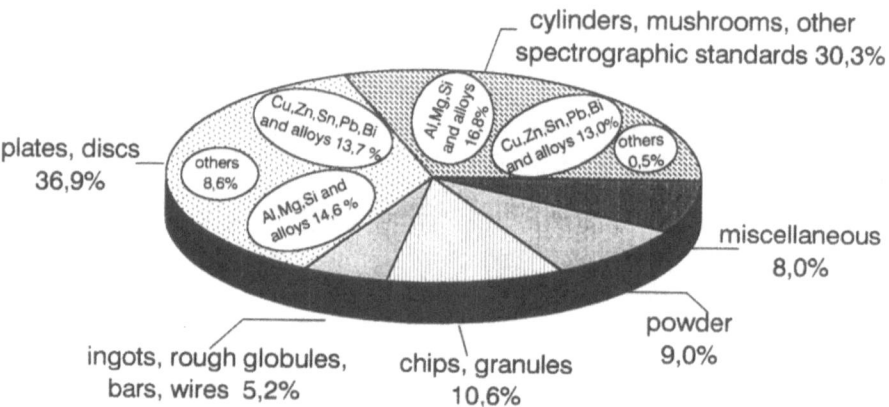

Fig. 4.4. Relative frequency of different forms of reference materials used for analyses of non-ferrous materials. Data come from the COMAR database and are based on about 3400 entries

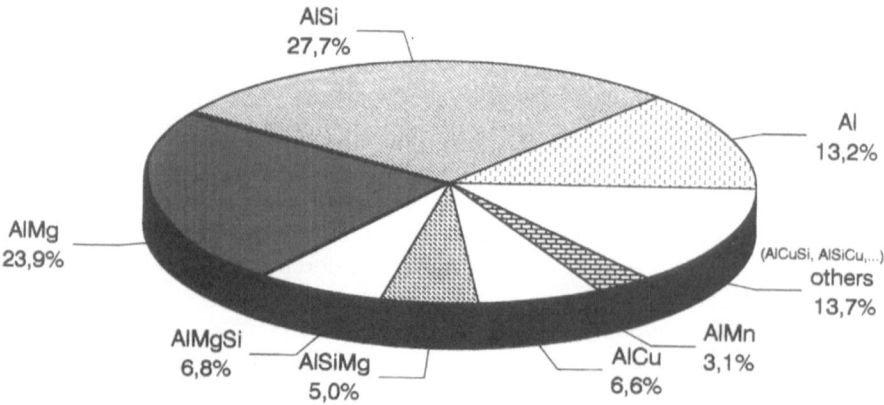

Fig. 4.5. Relative frequency of some important types of reference materials used for analyses of Al and Al alloys. Data come from the COMAR database and are based on about 900 entries

to more than 20%), trace contents (in most cases clearly below 1%) are certified for several analytes: Fe, Cu, Mn, Mg, Cr, Ni, Zn, Be, Ga, Pb, Sb, Sn, V, Ti and others. Additionally, in some of the reference materials, the elements B, Ca, Na, P and Sr are certified at lower trace concentration levels.

- In second place with almost one quarter of the group are the aluminium–magnesium alloys. The contents of the minor constituent Mg (a few %) and the trace contents of Si and of the elements mentioned above are certified. For some RMs certified values for Cd, Hg and Li at low concentration levels are also given.

- Approximately 13% in the considered category are CRMs for the analysis of pure aluminium and of non-alloyed aluminium. Certified analytes are similar to those of the samples mentioned above. The contents depend on the material type and the analyte and are at a level of some mg/kg up to about 0.1%.

- The number of CRMs that can be assigned to different classes of other alloys such as AlMgSi, AlSiMg, AlCu or AlMn is distinctly smaller as can be seen in Fig. 4.5. Further classes of alloy types, such as AlCuSi or AlSiCu have been summarized in one last segment.

There exist reference materials for most of the mentioned classes of alloy types, certified by multinational, national or national representative institutions such as:

- IRMM (Geel), European,
- NIST, USA,
- BAM, Germany (in cooperation with GDMB),
- Bureau of Analysed Samples Ltd., United Kingdom.

However, it is not possible to fulfill the demands for the whole range of alloy types by CRMs from public institutions. The producers of the largest number of reference materials of each of the discussed material groups are not the state institutions but leading producers of Al and Al alloys themselves. These are:

- Aluminium Pechiney, France,
- VAW aluminium AG, Germany,
- Alusuisse-Lonza Services AG, Switzerland.

If discrepancies arise between CRMs from different commercial producers, in some cases CRMs produced by independent, neutral public institutions not involved in the production of aluminium materials themselves may be used as a reference. To this end, in many cases only one CRM of a special type of alloy is sufficient because discrepancies between the different calibration curves can often be expressed as a constant factor over the whole calibration range.

Copper and Copper Alloys. Pb, Sn, Sb and Alloys

To a still greater extent than for aluminium alloys, it is difficult to give an unambiguous and useful classification for the groups referring to these materials. The following statements aim to give a summary of the assignment and distribution of RMs within classes (groups) of materials that are sometimes divided arbitrarily. As with the CRMs for Al and Al alloys, not only those RMs with given uncertainties are taken into account, but also RMs for which only the element contents are given without any uncertainty estimation.

Figure 4.6 shows the distribution of 910 different CRMs listed in the COMAR database for the analysis of Cu, Zn, Pb, Sn, Bi and their alloys as a rough grouping. The following conclusions may be drawn:

- Most of the samples, approximately one quarter, are RMs of CuZn-type. All samples with Cu contents more than 40% and with Zn contents from 1 up to 45%, Sn contents below 0.7% and Ni contents below 7% were assigned to this group. Thus, besides ordinary brass alloys based on Cu and Zn, other material types are also included, like red brass, gilding metal and some kinds of special brass (such as leaded brass or silicon brass). In most cases, the Zn content is certified to high accuracy, and sometimes also the Cu content. Other certified elements with contents of some tenth of % are typically Pb, Zn, Fe, Ni, Mn, Al and Si. Some special elements like P are also certified at lower concentrations. For the elements Pb, Fe, Ni, Mn, Al and Si clearly higher contents of a few % are certified for some types of material. The main producers of this group of RMs are EZKS Giprotsvetmetobrabotka, Russia, the British Bureau of Analysed Samples Ltd., U.K., and MBH Analytical Ltd., U.K. as well as NIST, USA, CTIF/Techlab, France, Japan Brass Makers, BAM, Germany and others.

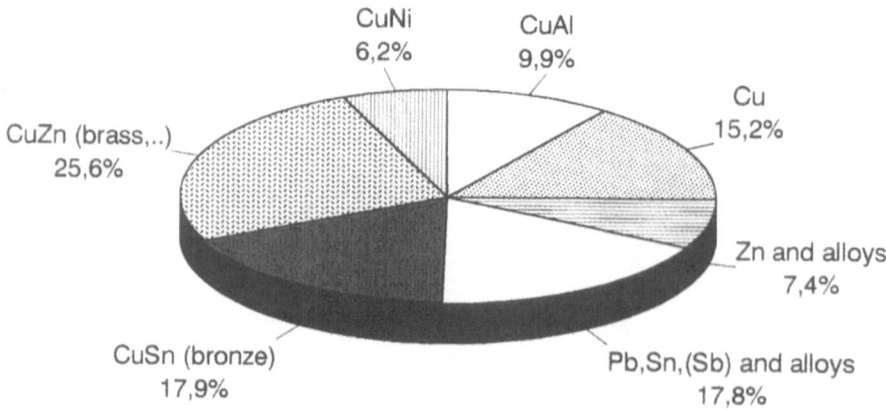

Fig. 4.6. Relative frequency of some important types of reference materials used for analyses of Cu, Zn, Sn, Pb, and alloys. Data come from the COMAR database and are based on about 910 entries

- Another important group with almost a fifth of all the materials in the discussed category are RMs of CuSn-type. This includes materials with high copper content, Sn-contents higher than 0.5% and Zn contents below 2.5%. Thus, Sn bronzes, special bronzes (like Si bronzes or leaded bronzes) are included in this group. Certified values are given for the Sn and sometimes also for the Cu content which was in some cases calculated as the difference of the sum of the other elements contained in the material to 100%. Typical traces like Pb, Zn, Ni, S, P, Sb, Fe, Si, Mn, Bi and Al are also certified in many of the RMs. Depending on the type of material, Pb, Zn, Ni and sometimes also other elements are contained and certified with higher contents of a few % (in some cases for lead at an even higher level). The main producers of RMs for the analysis of bronzes are CTIF/Techlab, France, EZKS Giprotsvetmetobrabotka, Russia, MBH Analytical Ltd., U.K., the British Bureau of Analysed Samples Ltd. and various Chinese producers.
- About 15% of the RMs have high pure, pure or very low alloyed copper qualities. The main part of this group consists in the purer copper materials sometimes with very low contents certified for a lot of trace elements at a level of a few or a few tenths of mg/kg. Unfortunately, some producers do not generally give any uncertainties for the certified values of their RMs. For low alloyed materials, Mg, Ni or Cr contents are certified in the % region. There exist 15 producers for this group of CRMs. The main producers/distributors are Breitländer, Germany, EZKS Giprotsvetmetobrabotka, Russia, and Dimitar Blagoev Metallurgical Enterprise, Bulgaria. But state or similar institutions like IRMM, Geel, (Europe), NIST, USA, BAM, Germany and the British Bureau of Analysed Samples Ltd. also produce RMs of the type mentioned, some of them with a high metrological quality or for the analysis of very specific materials.
- Approximately 10% of the copper-based RMs are materials of CuAl-type here defined by Cu contents of more than 40% and additionally Al contents of more than 7%. Materials like cupro aluminium, aluminium bronzes, MnAl-bronze and AlFeNiMn-bronze are summarized in this group. Besides trace element contents and the Al content (often about 10%), the copper content as well as the contents of Fe, Zn and Ni (a few %) are often certified.
- Some further materials were assigned to a group of the CuNi materials (6.2% of the investigated CRMs) with Cu contents > 45% and Ni contents > 7%. This group contains alloys like cupro nickel, Ni silver and CuMn-brass.
- In the group of non-cuprous materials the most frequent part (17.8%) is made up of CRMs for Pb, Zn and (to a lesser extent) Sb and their alloys. This group is dominated on the one hand by pure lead and low alloyed lead with less than 5% alloying elements (mostly Sb and Ag). On the other hand the SnPb solders and SnPbSb solders with higher contents of Sn,

Pb or Sb, respectively, constitute a second extended part of this group. Besides the main components, several trace elements are certified in most cases, some of them with very low contents. The most important producers/distributors are Dimitar Blagoev Metallurgical Enterprise, Bulgaria, Metaleurop Recherche, France and MBH Analytical Ltd., U.K. However, state or similar institutions like the British Bureau of Analysed Samples Ltd., the NIST, USA and the European IRMM produce CRMs of this type, too.

- A second group of non-cuprous materials are less frequent RMs (7.4%) based on Zn and Zn alloys. This group contains pure Zn with low concentrations of trace elements (typical are Al, In, Pb, Sn, Tl, Cd, Cu, Fe) in the mg/kg region and lower. Producers of this type of material are the European IRMM and Dimitar Blagoev Metallurgical Enterprise, Bulgaria. Additionally, there are RMs for Zn alloys, e.g. ZnAl and ZnAlCu, produced by IRMM, MBH Analytical Ltd., U.K. and the Primary Station of Chemical Metrology, China. Certified values are given for the alloying elements in the % region and for trace impurities in the low mg/kg region (but for Mg: hundreds of mg/kg).

Other Non-Ferrous CRMs

The CRMs not described under the two preceding subheadings are mainly distributed over the following groups of materials (Fig. 4.3):

- Ni, Co, Cr and refractory metals (8.5%) in which base alloys and special alloys with main components of Ni, Cr and Co are most frequent (main producers/distributors are: MBH Analytical Ltd., United Kingdom, EZKS Giprotsvetmetobrabotka, Russia, and NIST, USA).
- Rare earth, Th, U and the transuranic elements (6.4%). This group is dominated by CRMs with certified values for the isotopic composition of U and by CRMs for U metal as well as for U and Th ores. Mg(Ag)-rare earth alloys also belong to this category (main producers/distributors are: Techsnabexport, Uranservice, Russia; International Atomic Energy Agency (IAEA), Vienna, Austria; IRMM, Geel, Belgium; CEA, France and different Chinese producers).
- Pure metals (5.9%), partly distributed only with an indicative certificate of the producer and in most cases also assigned to other groups of materials, such as the Al or Cu group or the precious metals (main producers: Shenyang Nonferrous Metal Working, China; MBH Analytical Ltd., United Kingdom).

In addition, there are further reference materials summarized in the categories precious metals and alloys, raw materials and by-products or Ti, V and alloys. Each of these categories contains approximately 2–3% of the non-ferrous RMs.

Finally it should be noted that the classification of reference materials made in Sect. 4.4.2 has only an informal character because of the wide variety of materials. Owing to the different metrological qualities of CRMs for non-ferrous materials, each CRM should be checked not only with regard to the price, form, producer, composition, etc., but also with regard to the quality of the certified values and their given uncertainties.

Some Details on BAM CRMs

Finally some mention should be made to the certification of non-ferrous metals prepared under the auspices of a state institution, the Federal Institute for Materials Research and Testing (BAM), carried out in collaboration with the Commitee of Chemists of the GDMB (Gesellschaft für Bergbau, Metal-lurgie, Rohstoff- und Umwelttechnik). Both the public institution and the industrial association have to decide on the chemical content and elements of future reference materials. BAM buys the raw material which is normally manufactured by one of the industrial partners specially for this purpose. After preparation of the samples and pilot tests (mechanical preparation, e.g. grinding, feasibility study, homogeneity test) BAM sends the material to 10 to 15 experienced and qualified laboratories participating in the certification interlaboratory comparison (round robin). Each laboratory is requested to determine the content of each relevant element a total of six times. If possible they are asked to use analytical procedures based on several different methods in order to uncover and avoid typical method-related errors. Each analytical procedure must only be calibrated with pure substances or with solutions (or fused digested materials) prepared from them in order to establish a metrological traceability. Calibration with other CRMs is not allowed.

After gathering results, a survey and a statistical evaluation of the data from the laboratories is carried out by BAM with respect to their distribution, the calculation of laboratory and total standard deviations and confidence intervals as well as to the identification of any outlying values (Grubbs, Cochran). The corresponding working groups of the Committee of Chemists at the GDMB ("Copper", "Aluminium", "Lead", "Tin", "Precious Metals" or "Special Materials") and also the Committee of Certification at the BAM then discuss the results. After confirmation of the contents as certified values the respective certificate is written and the samples made ready for sale.

4.4.3 Reference Materials for the Glass Industry

In the glass industry, RMs are used mainly for chemical analysis in order to determine the major, minor and trace amount components in glasses, glass ceramics, raw materials and materials connected with environmental protection.

In the glass industry and particularly in the specialty glass industry, glasses of various glass types serve as RMs. They fulfill various requirements regarding the certified elements and elemental contents and also the uncertainty budget. For the utility glasses of everyday life, like flat glass and crystal glass, the requirements regarding the determination of contents of minor and trace amount components are lower than for laboratory, container and pharmaceutical glass packaging. In the latter, an important role is played by elements which are crucial in the domain of life sciences and their determination must already be carried out with high accuracy at low concentrations.

For optical glasses, the detection of heavy metal contents with specific light absorption properties (e.g. Cr, Co, Mn, Fe) is of special importance, because these elements strongly influence the utility value. Therefore a whole series of reference materials with different matrices and elemental contents has been developed, corresponding to the various requirements encountered in the chemical analysis of glasses.

RMs are used in various forms. For wet chemical analyses, powder materials are available, while for XRF, which is often applied in the glass industry, compact samples are normally needed.

In the glass industry, one has the following types of glasses, for the analysis of which corresponding RMs must be used:

- soda-lime glass (flat glass, container glass);
- borosilicate glass (laboratory glass, pharmaceutical packaging);
- crystal glass (lead crystal glass, crystal glass with variable PbO contents or without PbO);
- television glass (panel glass, funnel glass, neck glass);
- glass ceramics;
- optical glasses;
- ophthalmic glasses.

For the glass industry, two kinds of reference materials are important:

- primary or superpure substances,
- certified matrix reference materials.

Primary or superpure substances are highly pure, whenever possible one-phase substances, the elemental content of which is known so precisely that it can be used for calibrating the measuring instruments or for controlling analytical results. The analyte content in these certified reference materials must have a significantly smaller uncertainty interval than required by the repeatabilty of the analytical method.

With regard to certified matrix reference materials, wet chemical analytical methods usually require only one reference sample, adapted to the glass matrix under scrutiny. For different glass types and raw materials, numerous certified RMs are available (Tables 4.4 and 4.5).

For instrumental analytical methods, like XRF or laser ICP-MS, reference materials must be used for calibration, with largely adapted glass matrix. Such reference materials can be molten from certified primary substances, if the errors due to vaporization are negligibly small or if powder compacts can easily be prepared. In the latter case, there must be a rigorous homogeneity test before applying the material. In addition, numerous commercial RMs are available in compact form with a broad range of elements and a broad concentration range (Table 4.6).

It is known from experience that the wet chemical investigation of molten reference glasses or of commercially available matrix RMs both yield results which are excellently comparable, e.g. between values determined by means of wet chemical methods and results from XRF. XRF can also be used for analyzing samples digested by fusion and hence for calibration using pure substances as starting materials.

The sources of supply of RMs are brought together in Table 4.7.

4.4.4 Reference Materials for Ceramics Analysis

Ceramics come with a wide range of compositions and exhibit a correspondingly wide range of applications as materials with specific mechanical, optical, electrical, thermic and other attributes.

The raw materials for modern specialised ceramics are metals and compounds which often show high melting points and high thermic stability. The specific requirements for chemical analysis follow from this fact. Such elements are found in the subgroups 4 up to 6 of the periodic system of elements, especially the carbides, oxides, borides and nitrides of titanium, zirconium, niobium, tantalum, molybdenum and tungsten. In the preliminary materials for high tech ceramics, aluminium, silicon and boron are found as matrix elements. Typical high technology ceramics are aluminium oxide (Al_2O_3), silicon nitride (Si_3N_4), aluminium nitride (AlN), titanium boride (TiB_2), boron carbide (B_4C), zirconium oxide (ZrO_2), zirconium-stabilized yttrium oxide (ZrO_2/Y_2O_3) and even more. The early development of these products was assisted by the evident possibilities of powder metallurgy and the corresponding analytical methods of determination.

For production of metals in the form of powders and inorganic compounds ores are normally used. Today many special metals are produced in recycling processes using scraps and other waste products, thus helping to save resources. The production process with several steps (raw materials, salts, oxides, metals, etc.) has to be controlled analytically and continuously.

For the qualitative and quantitative determination of the elements in ceramics and their educts today, instrumental methods are used in most cases. Therefore a suitable sample preparation (decomposition) and a calibration of the method is required to correlate the measured signal of the instrument with the mass fraction of the element to be determined. The calibration process ensures the traceability of the measured results. In order to obtain identical species in the solutions of samples and calibration standards, the sample is brought into solution by digestion. For the calibration, defined pure substances with known contamination levels are used. By weighing these pure

Table 4.4. Selected RMs for different types of glasses

Designation	Al_2O_3	As_2O_3	B_2O_3	BaO	CaO	CeO_2	CdO	Cr_2O_3	F	Fe_2O_3	GeO_2	K_2O	Li_2O	MnO	Source
Standard glasses (Soda-lime)															
Standard glass I of DGG	1.23				6.73					0.191		0.38			A
Standard glass No. 6	1.70	0.30			9.97					0.034		< 0.1			B
Glass SRM 621	2.76			0.12	10.71					0.040		2.01			D
Soda-lime-glass JCRM 1102	1.58				7.22					0.094		0.76			G
Flat glass SV RM I	0.61		(0.014)		8.61			(0.0009)		0.042		0.30			C
Bottle glass, green SVRM III	2.13		0.061		9.89			0.11		0.290		0.94			C
Bottle glass, brown SVRM IV	1.80		(0.025)		7.33			0.004		0.300		0.70			C
Borosilicate glasses															
Standard glass No. 2	2.49		12.86		0.20					0.075		0.23			B
SRM 1411	5.68		10.94	5.00	2.18					0.050		2.97			D
Lead crystal glasses															
CRM 126 A	0.128		1.04		1.03					0.0055		10.00	0.495		E
Standard glass No. 8	0.05	0.32	0.36		< 0.02					0.010		11.85			B
SRM 89	0.18	0.34		1.40	0.21					0.049		8.40		0.088	D
Opal glasses															
Standard glass No. 4	3.02			0.19	4.24				4.96	0.099		0.57			B
SRM 91	6.01				10.49				5.73	0.079		3.24			D
Glasses for microanalysis (fibres)															
K-493	0.2		0.14			0.68				0.32			0.001		D
K-491	0.16		0.11			0.56				0.26			0.001		D
K-411					15.47					16.02	37.98				
Other glasses															
SRM 1412 (Multicomponent)	7.52		4.53	4.67	4.53		4.38			0.031		4.14	4.50		D
SRM 92 (Low boron)			0.70		8.30							0.6			D
A FF 11 (Silica brick)	4.14				0.10					0.49		0.46			C

Table 4.4. (continued)

Designation	MgO	Na₂O	P₂O₅	PbO	SO₃	SrO	Sb₂O₃	SiO₂	TiO₂	Ta₂O₅	ZnO	ZrO₂	Source
Standard glasses (Soda-lime)													
Standard glass I of DGG	4.18	14.95			0.436			71.72	0.137				A
Standard glass No. 6	< 0.1	14.65			0.20			73.06	0.02				B
Glass SRM 621	0.27	12.76			0.13			71.13	0.14			0.007	D
Soda-lime-glass JCRM 1102	4.01	13.36	0.013		0.22			72.51	0.040			0.005	G
Flat glass SV RM I	3.99	13.50	(0.012)		0.24			72.40	0.037			(0.012)	C
Bottle glass, green SVRM III	2.82	13.30	(0.013)		0.15			70.10	0.047			(0.013)	C
Bottle glass, brown SVRM IV	4.74	13.42	(0.0087)		0.038			71.80	0.033			(0.0087)	C
Borosilicate glasses													
Standard glass No. 2		3.85						80.08	0.035				B
SRM 1411	0.33	10.14				0.09		58.04	0.02		3.85		D
Lead crystal glasses													
CRM 126 A	0.512	3.58		23.98			0.290	57.80			1.02		E
Standard glass No. 8	< 0.02	0.23		30.59				56.34	0.02				B
SRM 89	0.03	5.70	0.23	17.50				65.35	0.01				D
Opal glasses													
Standard glass No. 4	< 0.05	15.45			< 0.05			69.49	0.041		3.28		B
SRM 91	0.008	8.47		0.10				67.50	0.019		0.08	0.009	D
Glasses for microanalysis (fibres)													
K-493				69.08				27.89	0.32	0.88		0.49	D
K-491				59.35				0.19	0.26	0.72		0.40	D
K-411	14.67					4.55		54.30					
Other glasses													
SRM 1412 (Multicomponent)	4.69	4.69		4.4				42.38			4.48		D
SRM 92 (Low boron)	0.1	13.1						75.00			0.2		D
A FF 11 (Silica brick)	0.11	0.04						92.24	0.40				C

Table 4.5. Selected RMs for raw materials and environmental samples

RM	Nr.
Limestone	SRM 1c
Limestone dolomitic	SRM 88b
K feldspar	SRM 70a
Na feldspar	SRM 99a
Clay, flint	SRM 97b
Clay, plastic	SRM 98b
Clay, brick	SRM 679
Glass-sand	SRM 81a
Glass-sand (low iron)	SRM 165a
Glass-sand (high alumina)	SRM 1413

(Source of supply: D)

RM	Nr.
Kryolith	V K01-2 bis V K22-2
Bauxite	V B01-2 bis V B16-1
Syenite	S 201 a
Zr silicate	S 204 a
Na feldspar	IP 72
Fluorite	GB 07250 bis GB 07254
Limestone	GB 07214 bis GB 07217
Limestone	VS 3192-85 bis VS 3193-85

(Source of supply: C)

RM	Nr.
Soils	CRM 141, 142, 143
Sediments	CRM 277, 280, 320
Phospate rock	CRM 032
Corundum	RM 300
Mullite	RM 301

(Source of Supply: E)

substances, traceability to SI units is ensured if the purity of the substances is well characterized. For testing whether the digestion procedure is complete and free of losses, there is a preferred reference method, which is independent of sample digestion, physical measurement and calibration.

Regarding the diversity of new inorganic materials in the science of ceramic materials, and in the electronic and optics industries today, no industrial laboratory is able to make such reference methods available for a great

Table 4.6. Selected RMs for X-ray fluorescence analysis (source: C)

	RM					
Element	2T U39	BR 2/L	BR 8/L	2T U16A	2T U34	2T U22/2
Al_2O_3	1.0	4.680	8.070	1.0		7.0
As_2O_3		0.500	0.420	1.5		
Ag_2O						0.5
BaO		2.900			7.8	
B_2O_3		13.800	7.470	25.0		8.0
Bi_2O_3						0.5
Br						0.1
CaO	6.0	3.93	0.020		6.5	10.0
CdO		0.430				
CeO_2		1.010		1.0		
Cr_2O_3					0.25	
CuO					0.1	
F			4.020			
Fe_2O_3	0.05	0.100	0.020		0.03	
GeO_2				1.5		
I						0.1
K_2O	0.5	3.120	20.100	5.0	5.4	
La_2O_3		1.000		1.000		
MgO	4.0	6.730	0.010	10.0		6.5
MoO_3				1.0		
Na_2O	15.0	6.820	0.070	10.0	9.3	16.0
Nb_2O_5				0.050		1.0
Nd_2O_3						0.5
NiO						2.0
PbO		3.040	1.200			
P_2O_5		1.260				0.3
Pr_2O_3						0.5
SO_3	0.35	0.120				
Sb_2O_3		0.500			0.35	
SiO_2	73.0	44.600	49.400	38.0	69.7	45.0
SnO				0.5		
SrO		0.700				
Ta_2O_5				0.1		
TiO_2		0.990	9.200			
V_2O_5				3.0		
WO_3				1.0		
ZnO		2.670				1.0
ZrO_2		0.500				

Table 4.7. Source of supply for RMs

Source A:	Deutsche Glastechnische Gesellschaft DGG Mendelssohnstr. 75–77 60325 Frankfurt am Main, Germany

Source B:	Society of Glass Technology H.R.T. Labortechnik GmbH Dieselstr. 6 80993 Kirchheim bei München, Germany

Source C:	Breitländer Eichproben und Labormaterial GmbH Postfach 80 46 59067 Hamm, Germany

Source D:	Office of Standard Reference Material V.S. Department of Commerce National Institute of Standards and Technology NIST Rm 8311, Chemistry Bldg. Gaithersburg, MD 20899, USA

Source E:	Community Bureau of Reference (BCR) Commision of the European Communities Directorate General for Science Research Development 200 Rue de la Roi 1049 Brussels, Belgium

number of elements. Reference materials are therefore needed and used when available to check routine analyses in industrial laboratories.

Selfmade reference materials, so-called "in-house standards", are usually used to solve this problem, because not enough certified RMs are available.

The following tables give a selection of a few available ceramic RMs. The element contents are often not certified, i.e. values are given for information only (indicative values). Tables 4.8–10 include oxidic reference materials, and lattice spacings determined by X-ray-diffraction are also given in the last table. Tables 4.11 and 4.12 contain carbidic reference materials with various compositions and from several producers. Table 4.13 shows the composition of a mixed oxide which is suitable for checking an analytical method.

In 2000 BAM will terminate the production of certified reference materials (Si_3N_4 and SiC) on a high metrological level with several certified metallic trace elements. Beyond this, the content of N and C and the α/β-ratio of phases will be certified for Si_3N_4.

Table 4.8. Standard reference materials (SRMs) for selected oxides (source: NIST)

Catalogue Number	SRM76a	SRM77a	SRM78a	SRM154b
Material	Burnt refractory (Al_2O_3-40%)	Burnt refractory (Al_2O_3-60%)	Burnt refractory (Al_2O_3-70%)	Titanium dioxide
Certified constituent	Certified value [weight %]			
Al_2O_3	38.7	60.2	71.7	
CaO	0.22	0.05	0.11	($\sim$ 0.01)
Fe_2O_3	1.60	1.00	1.2	(0.006)
Li_2O	0.042	0.025	0.12	
MgO	0.52	0.38	0.70	($\sim$ 0.01)
P_2O_5	0.120	0.092	1.3	(0.04)
K_2O	1.33	0.090	1.22	
SiO_2	54.9	35.0	19.4	(0.01)
Na_2O	0.07	0.037	0.078	
SrO	0.037	0.009	0.25	
TiO_2	2.03	2.66	3.22	99.74
L.O.I.	(0.34)	(0.22)	(0.42)	

Table 4.9. Reference material for high purity alumina (Japan. Ceramic Soc., ICRM) (cf. COMAR)

Content [%]	R 0 31 [%]	R 0 32 [%]	R 0 33 [%]
SiO_2	0.0009[a]	0.015[b]	0.095[a]
TiO_2	0.0000[b]	0.002[b]	0.004[b]
Fe_2O_3	0.0002[b]	0.013[b]	0.015[b]
CaO	0.0001[b]	0.034[a]	0.020[b]
MgO	0.0000[b]	0.001[b]	0.001[b]
Na_2O	0.0019[a]	0.343[a]	0.064[a]
K_2O	0.0003[b]	0.003[b]	0.003[b]
SO_3	0.017[a]		

[a] Certified value.
[b] Indicative value.

Table 4.10. RMs of α-alumina and mullite: chemical composition and lattice spacing (source: IRMM/BCR)

Corondum (β-alumina) RM 300 (not certified)

High crystallinity.
β-alumina content about 6 mg/g. No other phase detected.

Impurities [g/kg]:

SiO_2 < 5
Fe_2O_3 < 1
CaO < 1
MgO < 0.5
Na_2O ≈ 3
K_2O < 0.5
TiO_2 < 0.5

Mullite (3Al$_2$O$_3$·2SiO$_2$) RM 301 (not certified)

High crystallinity.
Vitreous phase 0.03 g/g. No other phase detected.

Impurities [g/kg]:
Fe_2O_3 < 2
CaO < 1.2
MgO < 0.5
Na_2O < 1
K_2O < 0.5
TiO_2 < 0.5

α-Alumina RM 300			Mullite RM 301		
Reflection	Lattice spacing [nm]	Relative intensity	Reflection	Lattice spacing [nm]	Relative intensity
[012]	0,3479	0,73	[110]	0,5382	0,50
[110]	0,2381	0,40	[210]	0,3390	1,0
[113]	0,2086	1,0	[220]	0,2695	0,40
[024]	0,1741	0,43	[121]	0,2206	0,59
[116]	0,1602	0,79	[331]	0,1524	0,36

The reference materials are available in granular form. The user must grind them following the same procedure as for the samples under investigation. However, care should be taken to avoid overgrinding.

Table 4.11. Standard reference materials (SRMs) of selected carbides (source: NIST)

Catalogue Number	SRM887	SRM888	SRM889	SRM112b	SRM276b
Material	Cemented carbide (W83-Co10)	Cemented carbide (W64-Co25-Ti4)	Cemented carbide (W75-Co9-Ta5-Ti4)	Silicon carbide	Tungsten carbide
Certified constituent	Certified value [%]				
Aluminium				0.44	
Carbon (total)	(5.5)	(4.6)	(6.0)	29.43	6.10
Carbon (free)				0.26	(0.04)
Calcium				0.04	
Cobalt	10.35	24.7	9.50		
Iron				0.13	
Oxygen					(0.08)
Nirogen					(0.01)
Silicon carbide				97.37	
Tantalum		4.77	4.60		
Titanium			4.03		

Table 4.12. Reference material of nitrogen bearing SiC (BAS: BCS-CRM 359)

Element	Certified value [%]
C	23.46
Si	67.6
Al	0.118
Fe	0.175
Ca	0.108
Ti	0.022

Table 4.13. Standard reference material (SRM 699-NIST) of alumina for use in checking chemical methods of analysis and calibrating of instrumental analyses

Element	Certified value [%]	Uncertainty [%]
SiO_2	0.014	20
Fe_2O_3	0.013	8
Na_2O	0.59	2
CaO	0.036	6
ZnO	0.013	20
Ga_2O_3	0.010	20
MnO	0.0005	20
V_2O_5	0.0005	40
P_2O_5	0.0002	50
Cr_2O_3	0.0002	50
MgO	0.0006	40
Li_2O	0.002	50
BeO	0.0008[a]	
B_2O_3	< 0.001[a]	
ZrO_2	0.0002[a]	
CuO	0.005[a]	

[a] indicative content

4.4.5 Reference Materials for Particle Size Analysis

Disperse materials are of fundamental importance to a large number of industries [4.39–42]. Examples are those which manufacture or use catalysts, ceramics, adsorbents, building materials, drugs and other pharmaceutical products like tabletting agents, pigments and other additives for paints, inks, varnishes and so on. The properties of these particulate materials, such as reactivity, separation and sintering behavior, agglomeration, etc., depend critically on particle size and particle size distribution. Particle size dimensions measured can vary from the nanometer scale up to several millimeters.

The many techniques and devices used for off- and on-line analysis applied to measurement of these particulate properties require checking and calibration with a certified reference material [4.43].

A frequently encountered problem in the measurement of particle size parameters is that different results are obtained on ostensibly identical materials depending on which techniques are used. Different surface and size parameters are actually measured by these techniques and values can differ quite markedly.

It is thus necessary to describe the method used to measure particle size distribution, for example, gravitational sedimentation/balance method.

Another source of erroneous values is insufficient care taken when extracting small representative samples for measurement from bulk quantities of the material. The most frequent explanations involve different and sometimes erroneous applications of the detailed procedures. Samples for measurement must be carefully prepared, and measurement devices must be checked and calibrated. Certified reference materials are extremely valuable for this purpose [4.43].

When reference materials are used for particle size analysis the measurement results must be supplemented by information on the applied measurement technique, on suspending liquids and dispersing agents and on the entire procedure, indicating for instance the maximum solid content of a suspension, application of ultrasonic techniques for dispersing agglomerates, concentrations of dispersing agents, etc. The specifications regarding methods of preparation of samples for a number of measurement principles are laid down in various standards (DIN, CEN or ISO standards). As particle size can only be measured in certain particle size classes and as smaller and bigger particles outside these ranges are not detected, only reference materials can guarantee accurate instrument calibrations.

Specific Problems with Certification of Particle Size Reference Materials

The measurement of particle size or particle size distribution is an important analytical parameter for characterising a material and for monitoring and controlling chemical processes [4.44,45].

In general the viscosity of a dispersion of a solid material determines its properties. This is why particle size measurement is applied for both product characterisation and process control (e.g. crystallisation and polymerisation processes).

Product characterisation is the most important application of particle sizing techniques. For this purpose suitable dispersions have to be prepared (except for sieving) and these are then measured by different methods. Process monitoring and control systems in the production of emulsions and suspensions generally include sensors for on-line particle size analysis.

Methods for Particle Size Analysis (ISO DIS 9276) [4.46–48]

Various analytical principles are applied for particle sizing techniques which are based on specific interactions between radiation and particles and which require different methods for sample preparation and different methods for applying reference materials.

The following sizing methods other than sieving are used for particle size analysis:

Laser Diffraction Methods (ISO DIS 13320) [4.49]. This technique is also known by other names, viz.,

- Fraunhofer diffraction,
- near-forward light scattering,

- low-angle laser light scattering, applicable to particle sizes ranging from about 0.1 µm to 1500 µm.

A representative sample, dispersed at a sufficient concentration in a suitable liquid or air, is passed through the beam of a monochromatic laser light source. The light scattered by the particles at various angles in the near-forward direction is recorded by a multi element detector and numerical values relating to the scattering pattern passed to a computer memory.

The scattering pattern is then transformed using an adequate optical model and mathematical procedure, into a particle size distribution.

Small-Angle X-Ray Scattering Method. When a narrow beam of X-rays passes through a powder layer containing ultrafine particles, it will be dispersed at small angles around the incident beam due to electron scattering in the particles. The distribution of scattered intensity is closely related to the particle size distribution.

The method is suitable for all ultrafine powders containing particles in the size range of 1 nm to 300 nm. It is also applicable to organic or inorganic colloidal sols.

It should not, however, be used for:

- powders containing particles with far from spherical morphology, except in special cases;
- powders containing micro- and mesopores;
- mixtures of powders.

Gravitational Liquid Sedimentation Methods (ISO DIS 13317-1) [4.50]. Gravitational sedimentation particle size analysis methods are among the many methods in current use for determining the size distribution of numerous natural and artificial particulate materials typically in the range from 0.5 µm to 200 µm.

The gravitational sedimentation method includes the following measuring principles:

- pipette methods,
- the X-ray gravitational technique,
- the photosedimentation technique,
- sedimentation balance methods.

The method is applicable to powders made up of particles having the same density and of comparable shape.

Sedimentation techniques may be classified either as incremental or cumulative. Incremental methods are used to determine the concentration (or suspension density) in a thin layer at a known distance from the surface of the suspension. Cumulative methods are used to determine the rate at which solids settle out of the suspension.

In order to check the good working order and accuracy of the apparatus, it is recommended to analyse regularly certified reference materials with a range of particle size distributions. The size distribution of these reference materials is expressed in the form of cumulative curves given as equivalent Stokes diameters and should be comparable with those obtained from the gravitational sedimentation test method used.

Centrifugal Liquid Sedimentation Methods. Centrifugal sedimentation particle size analysis methods are among the many methods in current use for determining the size distribution of many natural and artificial particulate materials, typically in the sub-5 μm size range.

This method of determining the particle size distribution is applicable to powders which can be dispersed in liquids or powders present in slurry form. A positive density difference must exist between the discrete and continuous phases, although centrifugal photosedimentation can be used for emulsions where the droplets are less dense than the liquid in which they are dispersed.

No single method of size analysis can be specified to cover all of the many different types of material encountered, but it is possible to recommend procedures that may be applied to the majority of cases.

Ultrasonic Detection. Due to physical interactions particle size analysis by ultrasonic extinction is independent of the viscosity of a dispersion of the particles to be detected as long as the individual particles may move freely in the agglomeration. Analyses can be made in any liquid with high solid concentrations, making this method particularly suitable for online particle size analysis of suspensions and emulsions. Particle sizes in the range of 0.01 μm to 3000 μm at solid concentrations of 1 to 70 percent by volume may be analysed with this method [4.51].

Image Analysis Methods. Measurement of particle size distributions by microscopy methods (electron and optical microscopes) is apparently simple, but because only a small amount of sample is examined, considerable care has to be exercised in order to ensure that the analysis is representative of the bulk sample. Because of the diverse range of equipment and sample preparations available, it is not intended to give a prescriptive procedure where use of individual methods does not jeopardize the validity of the data. For calibration and traceability the equipment must be calibrated to convert pixels into a measure of units for the final results. The calibration procedure must include verification of the uniformity of the field of view. An essential requirement of the calibration procedure is that all measurements must be traceable back to the standard meter. This can be done by calibration of the image analysis equipment with a certified standard stage micrometer with certified monodisperse spherical particles.

Reference Materials for Particle Size Analysis

Table 4.14 shows standard reference materials (NIST) for evaluating and calibrating specific types of particle size measuring instruments, including light scattering, electrical zone flow-through counters, optical and scanning electron microscopes, sedimentation systems, and wire cloth sieving devices. SRM 659 consists of equiaxed silicon nitride particles with a minimal amount of large agglomerates; SRM 1978 consists of granular, irregular shaped zirconium oxide particles with a minimal amount of large agglomerates; SRM 1982 consists of spheroidal particles measured using scanning electron microscopy, laser scattering, and sieving. SRMs 1003b, 1004a, 1017b, 1018b and 1019b each consist of soda-lime glass beads covering a particular size distribution range. SRMs 1690, 1691, 1692 and 1963 are commercially manufactured latex particles in a water suspension. SRMs 1960 and 1961 (also called "space beads") are latex particles in a water suspension. SRM 1965 consists of two different groupings of the SRM 1960 particles.

A series of quartz reference materials (Table 4.15) has been developed with which measurements based upon different techniques of particle size distribution measurement can be compared or the setting up of individual instruments can be checked.

For each reference material the distribution is expressed as a curve of the cumulative mass of particle undersize versus particle size. In the case of particles of less than 90 μm diameter their size is expressed as the equivalent Stokes' diameter determined from the settling rate of the particles in a

Table 4.14. SRM for particle size distribution (NIST)

SRM	Type	Size range (μm), nominal
659	Silicon nitride	0.2 to 10
1003b	Glass beads	10 to 60 (600 to 325 mesh)
1004a	Glass beads	40 to 170 (325 to 100 mesh)
1017b	Glass beads	100 to 400 (140 to 45 mesh)
1018b	Glass beads	220 to 750 (60 to 25 mesh)
1019b	Glass beads	750 to 2450 (20 to 10 mesh)
1690	Polystyrene (0.5% in H_2O)	0.895
1691	Polystyrene (0.5% in H_2O)	0.269
1692	Polystyrene (0.25% in H_2O)	2.982
1960	Polystyrene (0.4% in H_2O)	9.89
1961	Polystyrene (0.5% in H_2O)	29.64
1963	Polystyrene (0.5% in H_2O)	0.1007
1965	Polystyrene	9.84 (hexagonal array)
		9.89 (unordered clusters)
1978	Zirconium oxide	0.33 to 2.19
1982	Zirconium oxide	10 to 150

Table 4.15. CRM for particle size distribution (IRMM, LGC)

Form of quartz	Certified property	Size range [μm]	Catalogue number IRMM (BCR)	LGC
Powder	Stokes' diameter	0.35–3.5	066	B0-66
Powder	Stokes' diameter	2.4–32	067	B0-67
Sand	Volume diameter	160–630	068	B0-68
Powder	Stokes' diameter	14–90	069	B0-69
Powder	Stokes' diameter	1.2–20	070	B0-70
Powder	Volume diameter	50–220	130	B1-30
Powder	Volume diameter	480–1800	131	B1-31
Gravel	Volume diameter	1400–5000	132	B1-32

Table 4.16. Latex spheres of certified size (LGC)

Description	Certified values (average diameter of the latex spheres)	Catalogue number
Nominal 2 μm latex (0.02% solids)	2.223 ± 0.013 μm	B1-65
Nominal 4.8 μm latex (0.2% solids)	4.821 ± 0.019 μm	B1-66
Nominal 9.6 μm latex (1.4% solids)	99.475 ± 0.018 μm	B1-67

viscous fluid. For larger particles the equivalent volume diameter determined by sieving was preferred.

Latex spheres of certified size (Table 4.16) can be used for the calibration of many instruments for particle size measurement and, in particular, for blood cell sizing instruments.

The spheres are suspended in an aqueous solution of stabilisers. In each sample the spheres have a very narrow distribution (99% of the spheres within ± 2% of the certified diameter). The sizes of the spheres have been selected to correspond approximately to the volume of platelets (thrombocytes), red blood cells (erythrocytes) and white blood cells (leucocytes) respectively. In haematological applications, precautions must be taken, as red blood cells are not rigid spheres and cannot therefore be compared with latex spheres. Users should first establish the correlation between the red cell volume determined by the standard manual method and their automated sizing instrument.

Liquid Suspending Media

The correct choice of suspension liquid and dispersing agents is extremely important when reference materials are applied for particle size analysis. Table 4.17 shows examples for suspending liquids and dispersing agents appli-

Table 4.17. Examples for suspending liquids and dispersing agents (ISO 13317)

Material	Suspending liquid	Dispersing agent (0.5 g/L to 1.0 g/L unless otherwise stated)
Active carbon	Isobutanol	
Alkali salts	Cyclohexanol	
Aluminium	Cyclohexanol Cyclohexanone Isopropanol (propan-2-ol) Paraffin oil	Lead naphthanate
Aluminium oxide	Water Water Water Water n-Butanol n-Butylamine Linseed oil: xylene	Sodium pyrophosphate Sodium polymetaphosphate Sodium tartrate Hydrochloric acid (pH 3)
Anhydrite (anhydrous calcium sulphate changes readily to gypsum with traces of water)	Methanol (anhydrous, acetone-free)	
Uranium dioxide (UO_2)	n-Butanol n-Butanol: Isopropanol Methanol Aqueous glycerol Isobutanol Water	Sodium polymetaphosphate
Zinc	Ethanol n-Butanol Isobutanol Cyclohexanone Cyclohexanol Water Glycol	Sodium polymetaphosphate Sodium polymetaphosphate Calcium chloride
Zinc oxide (pigment)	Water Water Water Aqueous glycerol	Sodium polymetaphosphate Sodium pyrophosphate Trisodium orthophosphate

Table 4.17. (continued)

Material	Suspending liquid	Dispersing agent (0.5 g/L to 1.0 g/L unless otherwise stated)
Zircon ($ZrSiO_4$)	Water Water:ethanol (1:1)	Sodium pyrophosphate
Zirconium	Isobutanol Methanol	 Hydrochloric acid (0.01 g/L)
Zirconium dioxide	Water Water Water Aqueous glycerol	Sodium polymetaphosphate Oleic acid Sodium pyrophosphate

cable for the method of gravitational sedimentation for example. Attention must be paid to making the right choice of viscosity. The viscosity of the suspending medium should have a value such that the largest particle to be measured takes at least 1 min to reach the measurement zone.

4.4.6 Reference Materials for Specific Surface Area and Pore Size Analysis

Porous and highly dispersed materials represent a specific state of solids and have numerous applications in research and industry [4.42,52–55]. Examples of application are catalysts, adsorbents, pigments, building materials, and pharmaceutical products. Porous materials are of great importance for the development and optimization of such processes as sintering, chromatographic separation, and catalyzed chemical reactions.

Specific Problems with the Certification of Porous Solids

The development and provision of porous CRMs according to the BCR guidelines makes an important contribution to the comparability of experimental results. In contrast to the certification of the element content of CRMs in which the procedure involves different independent methods of determining the conventionally true value, the certification of the values of specific surface area, pore size, and specific pore volume is method dependent. Depending on the method used for determination of the above mentioned quantities one obtains different results which are not necessarily comparable. Because of the various physical or chemical interaction mechanisms, for example of electrons, neutrons, gases or fluids, with a porous solid in the form of electron microscopy, neutron scattering, gas adsorption, or intrusion of a liquid, respectively, the determined quantities are different. Hence it is necessary to explain the method by which the quantity was measured, for example, specific surface area, measured by gas adsorption, mercury intrusion, or small angle scattering, etc. The results are only comparable if the quantities were measured by the same method and under the same conditions. In the case of measuring the specific surface area by gas adsorption, information must be given about the adsorptive used.

One of the most important methods for the determination of characteristic porosity properties is the analysis of gas adsorption isotherms (gas adsorption method [4.41]). This method is based on the weak interaction between an inert gas and the surface of a solid (physisorption). With mesoporous solids (pore sizes of 2–50 nm), the physisorption isotherms usually show hysteresis between the adsorption and the desorption branches. The hysteresis loop contains detailed information on the mesopores. Condensation and evaporation phenomena of the adsorptive gas are the reason for correlation between shape and position of hysteresis and the pore geometry [4.42]. The foundation of this phenomenon of capillary condensation is the Kelvin equation. By means of an appropriate model the isotherm gives the pore volume and the pore size distribution (as pore volume distribution). Unfortunately, the information about porosity is obtuse and heavily 'coded' in the isotherm shapes and this renders difficult a single unequivocal interpretation of adsorption behavior and modeling of porosity structure. Up to now, an all-embracing theory of physisorption for the entire p/p_0 range does not exist. For this reason standardized evaluation methods are available only for certain isotherm types of the IUPAC classification [4.56]. Moreover, because of the fractal character of porous solids, their specific surface area obtained from the monolayer capacity according to the BET model depends on the size of the probe molecules. The smaller the size of the probe molecule the higher is the specific surface area found. Furthermore, the comparability of results is complicated by various choices of specific experimental conditions (sample pretreatment, temperature and duration of heating, and degassing) as well as by the variety of available evaluation procedures.

The first CRMs available in this field were materials issued by BCR, LGC, and NIST with certified data concerning the specific surface area (Table 4.18).

The first reference materials of defined pore size distribution with certified data concerning their specific pore volume and mean pore radius were developed and issued by BAM. Some aspects of the development of these porous CRMs are described in the following.

Development of Porous Reference Materials for the Gas Adsorption Method

General Procedure. As in the case of other CRMs, the production of certified porous reference materials is also based on ISO guides 30–35 and

Table 4.18. CRMs for the specific surface area (method: multipoint BET)

CRM	Material	Certified value [m^2/g]	Producer
CRM 169	α-alumina	0.104	BCR
CRM 170	α-alumina	1.05	BCR
CRM 171	Alumina	2.95	BCR
CRM 172	Quartz	2.56	BCR
CRM 173	Rutile titania	8.23	BCR
CRM 175	Tungsten	0.181	BCR
LGC2101	Graphitised black carbon (Sterling FT-G 2700)	11.1	LGC
LGC2102	Graphitised black carbon (Vulcan 3-G 2700)	71.3	LGC
LGC2103	Non-porous silica (Silica TK800)	152.5	LGC
LGC2104	Meso-porous silica (Gasil silica)	258.0	LGC
LGC2105	α-alumina	2.09	LGC
LGC2106	α-alumina	0.229	LGC
LGC2107	α-alumina	0.0686	LGC
LGC2108	α-alumina	0.692	LGC
RM8570	Calcined kaolin	10.9	NIST
RM8571	Alumina	158.7	NIST
RM8572	Silica-alumina	291.2	NIST

the BCR-guidelines (BCR 48/93). Because of the above-mentioned restrictive conditions for the development of certified reference materials for pore analysis by means of gas adsorption, the materials to be certified must be selected according to a "suitable" isotherm type in conformity with the IUPAC classification [4.56]. With regard to the determination of the specific surface area of a porous material, the area occupied by the probe molecules is the decisive quantity for the traceability of calculated values. Therefore, the traceability required for certified reference materials can only be assured by applying strict standardized measurement and evaluation of the applied method (Fig. 4.7), DIN 66131–135 [4.57–61], ASTM 4404 [4.62].

The flow diagram for the certification procedure of a CRM with regard to its characteristic values of specific surface area, pore volume, and mean pore radius as a measure of the pore width is shown in Fig. 4.8.

For the area effectively occupied by a nitrogen molecule in a monolayer physisorbed at 77 K (molecular cross sectional area), the 1984 IUPAC recommendations give a value of 0.162 nm^2; this value has also been definitely fixed in the standard DIN 66 131 [4.57]

Some Experimental Details and Statistical Analysis of Data. Four materials, silicon dioxide, α-alumina, alumina type 60, and alumina type 150, were certified with respect to the above-mentioned characteristic values [4.63].

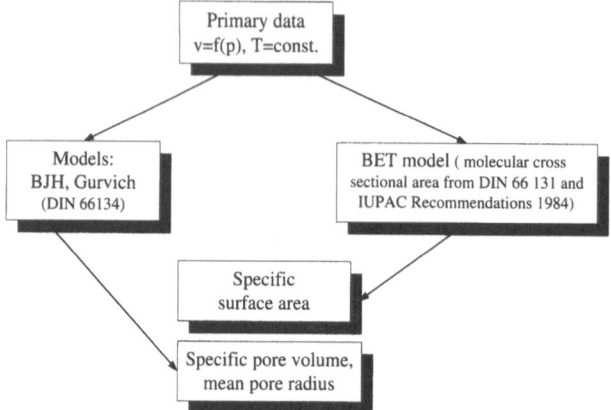

Fig. 4.7. Adherence to standardized procedures for assuring traceability [4.63]

For silicon dioxide, the gas used was krypton; for the other three materials nitrogen was used. For silicon dioxide and α-alumina, the specific surface area was certified; for alumina type 60 and alumina type 150 the specific pore volume and the pore radius were also certified. The specific surface area was determined according to the standard DIN 66 131 [4.57] (BET method) while the pore size distribution was determined according to BJH method, as described in DIN 66 134 [4.60] and ISO 9277. The results were found to be evaluation software dependent. Therefore it was important to evaluate the primary data (isotherm data) using the same software (called version 1 in Fig. 4.8) to obtain a minimum number of outliers and stragglers.

The statistical evaluation of quantities determined from the isotherms can be carried out according to DIN/ISO 5725 or according to BCR guidelines. Evaluation according to DIN/ISO 5725 presupposes, however, that the number of measured values is nearly the same for all laboratories, and that variances are homogeneous.

According to the BCR guidelines, "pooling of all individual data" (i.e. all sets of data produced by the various laboratories may be considered as samples from a single population of data and hence treated as one single set of data) is recommended only if the means and variances do not differ significantly. In the present case the means and variances of the laboratories were heterogeneous due to the use of different measurement apparatus, so that pooling was not allowed.

In the case of "no pooling", the BCR guidelines use a very simplified model for the calculation of the certified value. This means that laboratory mean values x_i are modeled and not the individual values. The laboratory means are modeled as the sum of the "true" value μ of the property to be certified, and the deviation (bias) $\Delta\mu_i$ of laboratory i, where the $\Delta\mu_i$ are considered to be statistically independent and have expectation zero.

$$x_i = \mu + \Delta\mu_i \qquad i = 1, \ldots, n \, (n- \text{ number of laboratories}) . \qquad (4.1)$$

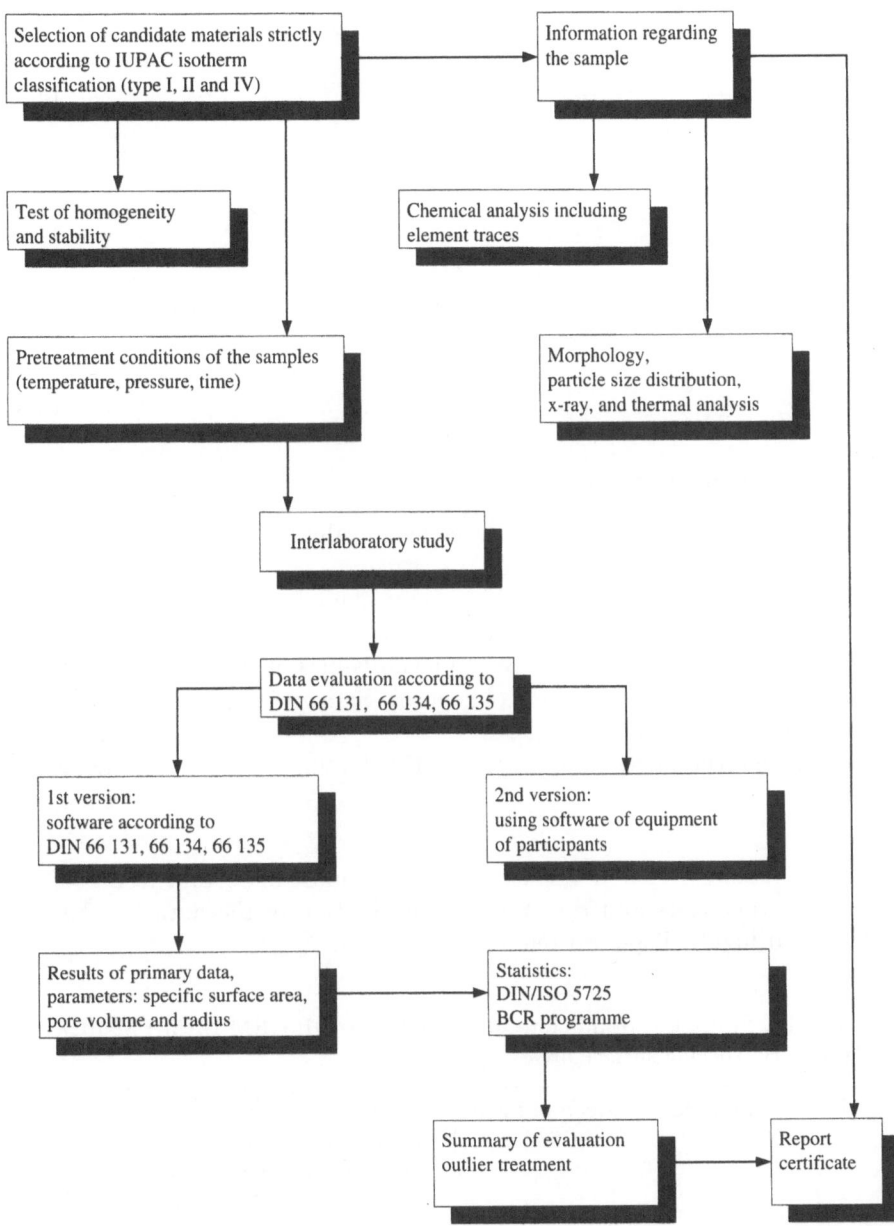

Fig. 4.8. Certification procedure of a CRM for the characteristic properties surface area, specific pore volume, and pore radius of highly dispersed and porous materials. Method: gas adsorption

Let $k_i \geq 2$ be the number of repeated measurements of laboratory i. The sample mean of laboratory i is used as an estimate for the laboratory mean

$$x_i = \frac{1}{k_i} \sum_j x_{ij} \,, \qquad i = 1, \ldots, n \,, \quad j = 1, \ldots, k_i \,, \qquad (4.2)$$

with a laboratory standard deviation s_i given by

$$s_i^2 = \frac{1}{(k_i - 1)} \sum_j (x_{ij} - x_i)^2 .$$ (4.3)

The "mean of means"

$$x = \frac{1}{n} \sum_i x_i , \qquad i = 1, \ldots, n ,$$ (4.4)

provides a reasonable estimate for μ and is used as the certified value. The standard deviation s of laboratory means is

$$s^2 = \frac{1}{(n - 1)} \sum_i (x_i - x)^2 , \qquad i = 1, \ldots, n .$$ (4.5)

Therefore, the procedure used was:

1. test for and eliminate outliers in the mean values of the laboratories using Dixon's method (1%, two-sided, iterative);
2. because of the mentioned heterogeneity, the laboratory dispersions were not subjected to an outlier test;
3. test for normal distribution of the mean values of the laboratories using the Lilliefors version of the Kolmogorov–Smirnov test;
4. determine the certified value;
5. determine the uncertainty of the certified value as a 95% confidence interval.

Results. Tables 4.19 and 4.20 summarize the most important results of the interlaboratory tests and the statistical evaluation of the data for the four porous and highly dispersed reference materials.

Table 4.19. Statistical results for CRM BAM-PM-101, CRM BAM-PM-102 certified property: specific surface area

CRM	Method	Mean of means (certified value)	Standard deviation of the mean of means [m^2/g]	Uncertainty 95%-confidence interval of mean of means [m^2/g]	Standard deviation of means [m^2/g]
CRM BAM-PM-101 material silica	BET (DIN 66 131), sorptive krypton	0.177 m^2/g (14 participating laboratories)	0.004	0.008	0.014
CRM BAM-PM-102 material α-alumina	BET (DIN 66 131) sorptive nitrogen	5.41 m^2/g (30 participating laboratories)	0.04	0.09	0.24

Table 4.20. Statistical results for CRM BAM-PM-103, CRM BAM-PM-104 certified properties: specific surface area, specific pore volume, mean pore radius, most frequent pore radius sorptive: nitrogen

Property	Method	Mean of means (certified value)	Uncertainty		
			Standard deviation of the mean of means	95% confidence interval of the mean of means	Standard deviation of means
Specific surface area [m²/g] 30 laboratories[a] 26 laboratories[b]	BET (DIN 66 131)	156.0[a] 79.8[b]	1.3[a] 0.4[b]	2.7[a] 0.8[b]	7.2[a] 2.0[b]
Specific pore volume [cm³/g] 24 laboratories[a] 22 laboratories[b]	Adsorption Gurvich (DIN 66 134)	0.250[a] 0.210[b]	0.002[a] 0.002[b]	0.004[a] 0.004[b]	0.008[a] 0.009[(b)]
Mean pore radius [nm] 24 laboratories[a] 22 laboratories[b]	$2V/S_{BET}$ (DIN 66 134)	3.18[a] 5.31[b]	0.02[a] 0.05[b]	0.03[a] 0.11[b]	0.08[(a)] 0.24[b]
Most frequent pore radius [nm] 26 laboratories[a] 25 laboratories[b]	BJH (DIN 66 134)	1.93[a] 3.23[b]	0.04[a] 0.05[b]	0.07[a] 0.10[b]	0.18[a] 0.23[b]

[a] CRM BAM-PM-103, material alumina type 60
[b] CRM BAM-PM-104, material alumina type 150

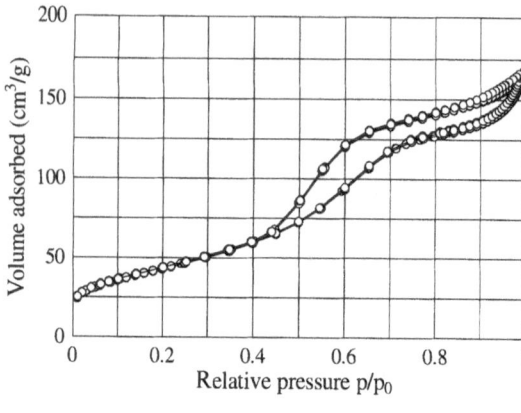

Fig. 4.9. Adsorption isotherms of nitrogen at 77 K on alumina type 60 results of single laboratory (9 measurements)

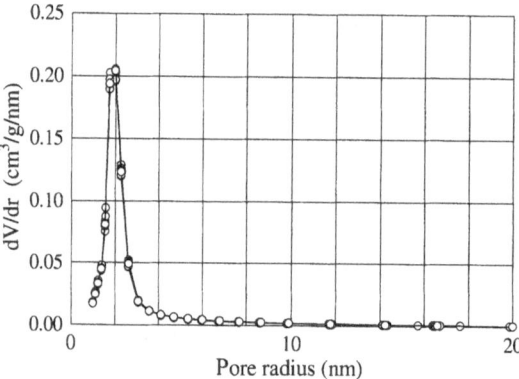

Fig. 4.10. Pore size distribution of alumina type 60 results of single laboratory (9 measurements)

To illustrate these tables, some results of the certification of the reference material CRM BAM-PM-103 are presented in Figs. 4.9 and 4.10. This substance with the manufacturer's designation alumina type 60 (mixture of transition aluminas) shows a N_2 isotherm (Fig. 4.9) of type IV according to the IUPAC classification. Figure 4.10 shows the pore size distribution as calculated by the BJH method.

Figure 4.11 represents an example of results of the statistical evaluation of a certified property of the porous and highly dispersed reference material CRM BAM-PM-103 (alumina type 60).

All of the above porous CRMs are mesoporous materials. Suitable candidates for microporous reference materials may be the group of zeolites.

An advantage of these microporous solids consists in the fact that they are crystalline and have a well-defined and regular pore network. On the other hand, there is not yet a universally accepted and standardized procedure for determining the size and distribution of micropores. Consequently, standardization work and measurements for the development of microporous CRMs must be carried out simultaneously.

For the characterization of macroporous materials with pore widths of more than 50 nm the method of mercury porosimetry must be applied. [Re-

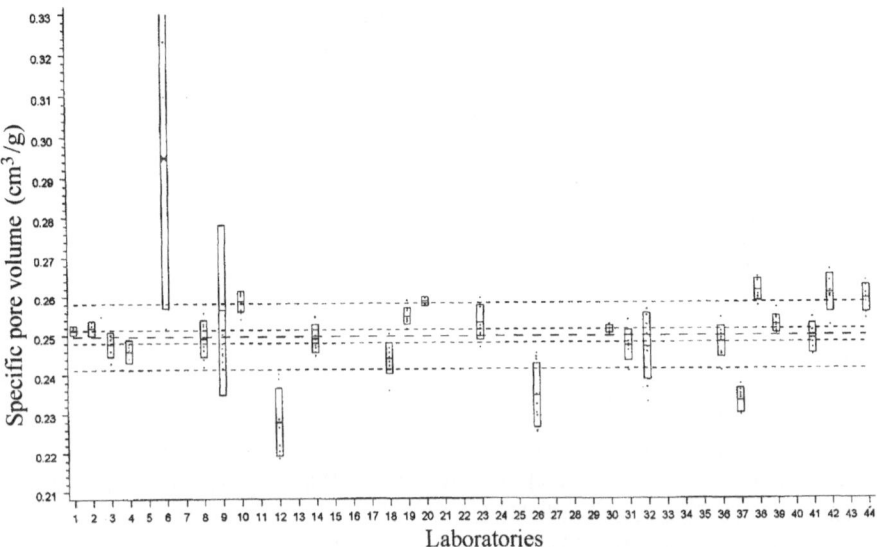

Fig. 4.11. Interlaboratory study: Specific pore volume (adsorption) of CRM BAM-PM-103 (alumina type 60). The *wide-dashed line* corresponds to the mean of means (certified value) with standard deviation (*narrow-dashed lines*). Laboratory mean values with standard deviation, individual values (*single points*), and standard deviation of means are also given. Crossed laboratory is an outlier

cently RMs for this method have been developed by BAM and can be ordered.] Further activities in the field of porous certified reference materials have to consider the development of microporous as well as macroporous reference materials. Such reference materials are in preparation in some current research projects at BAM.

Remarks on Use of Porous Reference Materials

Since the states of all solids depend more or less on their thermal pretreatment, it is of great importance to pay strict attention to the sample pretreatment procedures recommended for each RM (e.g. temperature and duration of the sample heating necessary for outgassing) when using all kinds of porous CRMs.

It is essential to observe the application conditions for a given reference material strictly, e.g. information on adsorptives (e.g. Kr, Ar, N_2 or other gases) and on adsorption temperatures (e.g. temperature of liquid nitrogen or of liquid argon). Only if these specifications are observed, can the values stated in the certificate be obtained or the instrument calibrated accurately. Otherwise erroneous results may be obtained. This also concerns the recommended purity of the gases used as adsorptives.

Therefore, it is essential to mention the measurement technique and the measurement conditions, e.g. "specific surface area determined according to gas adsorption by N_2 adsorption at the temperature of liquid nitrogen", etc. Unless these details are given, analytical results obtained by different laboratories cannot be compared.

According to the general requirement that calibrations should be based on similar materials to those analyzed, the development of reference materials for different classes of porous materials will become a task of growing importance over the next few years.

Acknowledgement

For discussions and contributions to Chap. 4, concerning examples of specific application, we would like to express thanks to the following colleagues: Dr. S. Recknagel/BAM (Sect. 4.1), Dr. K. Meier/BAM (parts of Sect. 4.2), L. Meckel/Schott Glas (Sect. 4.3.1), Dr. J. Peters/H. C. Starck (Sect. 4.3.2) as well as Dr. P. Klobes/BAM (Sect. 4.5), Dr. U. Beck, Dr. Th. Wirth/BAM (Sect. 4.3.2; part: glow discharge). We thank R. Pradel and S. Zimmer (BAM) for supporting our work with the COMAR database.

4.5   References

4.1. Rasberry S D (1989) The Role of Reference Materials in Measurement Quality Assurance. Proc. of ISCRM '89, Mai I5-18, Beijing, China, Certified Reference Materials, Intern. Academic Publ. (A. Pergamon - CNPIEC Joint Venture)

4.2. DAR-ATF, ATF/29a/97 (1997) Leitfaden zur Eignungsprüfung (Profiency Test) als Instrument der Akkreditierung im Prüfwesen, DAR-EM 17

4.3. Slickers K. (1992) Die automatische Atom-Emissions-Spektralanalyse, Buchvertrieb K. A. Slickers, Gießen, Germany

4.4. Hahn-Weinheimer P, Hirner A and Weber-Diefenbach K (1995) Röntgenfluoreszenzanalytische Methoden; Grundlagen und praktische Anwendung in den Geo-, Material- und Umweltwissenschaften, 2nd ed. Vieweg

4.5. Lachance G R and Claisse F (1995) Quantitative X-Ray Fluorescence Analysis; Theory and Application, John Wiley & Sons

4.6. Staats G and Noack S (1996) Qualitätssicherung in der Analytik; Die Rekonstitution; Eine Methode zur Optimierung der Richtigkeit von Analysen, Verlag Stahleisen, Düsseldorf, Germany

4.7. Hirschfeld D, Grallath E, Grau O, Gruner W and Schölz F (1997) Erzmetall **50**, 65

4.8. Boumans P W J M (Ed.) (1987) Inductively Coupled Plasma Emission Spectrometry, Part I: Methodology, Instrumentation and Performance; Part II: Applications and Fundamentals, John Wiley & Sons

4.9. Thompson M and Walsh J J (1989) Handbook of Inductively Coupled Plasma Spectrometry, Blackie Academic & Professional, 2nd ed.

4.10. Montaser A and Golightly D W (Eds.) (1992) Inductively Coupled Plasma in Analytical Atomic Spectrometry, 2nd ed., VCH Verlagsgesellschaft,

4.11. Winge R K, Fassel V A, Peterson V J and Floyd M A (1985) Inductively Coupled Plasma Atomic Emission Spectrometry; An Atlas of Spectral Information, Elsevier

4.12. Welz B and Sperling M (1998) Atomic Absorption Spectrometry, 3rd ed., Wiley-VCH

4.13. Butscher D J and Sneddon J A (1998) A Practical Guide to Graphite Furnace Atomic Absorption Spectrometry Chemical Analysis **149**, Wiley-Interscience

4.14. Date A R and Gray A L (1989) Applications of Inductively Coupled Plasma Mass Spectrometry, Blackie

4.15. Jarvis K E, Gray A L and Houk R S (1992) Handbook of Inductively Coupled Plasma Mass Spectrometry, Blackie

4.16. ISO/REMCO N420 06/1998: ILAC general requirements for the competence of reference material producers (final draft) (equivalent to the style and purpose of ISO/IEC Guide 25)

4.17. ISO/REMCO N425 03/1997: ILAC requirements for accreditation of certifiers of reference materials (draft)

4.18. Production and Use of Reference Materials Proc. Intern. Symp. held at the Bundesanstalt für Materialforschung und -prüfung (BAM) November 13–16 (1979)

4.19. Certified Reference Materials, Proceedings of ISCRM '89, May 15–18, 1989 Beijing, China, A Pergamon–CNPIEC Joint Venture

4.20. Breitkreutz K, Uttech R and Haedecke K (1988) Druckgesinterte Stähle als zertifiziertes Referenzmaterial für die Spektrometrie; Forschungsbericht der Bundesanstalt für Materialforschung und -prüfung (BAM), Wirtschaftsverlag NW, Bremerhaven

4.21. Breitkreutz K und Uttech R (1985) Metall **39**, 336

4.22. ISO Guide 30 (1992) Terms and definitions used in connection with reference materials

4.23. ISO Guide 31 (1996) Contents of certificates of reference materials

4.24. ISO Guide 32 (1997) Calibration of chemical analysis and use of certified reference materials

4.25. ISO Guide 33 (1989) Uses of certified reference materials (under revision)

4.26. ISO Guide 34 (1996) Quality system requirements for reference material producers

4.27. ISO Guide 35 (1989) Certification of reference materials – general and statistical principles (under revision)

4.28. Rasberry S D, Delve M, Mandry P (Eds.) (1992) A Survey of Organizations and Laboratories Manufacturing, Supplying or Using Reference Materials for Environmental Measurement, pp 110–111, GSF(Research Centre for Environment and Health) Neuherberg near Munich

4.29. Guidelines for the Production and Certifcation of BCR Reference Materials; Part A: Recommendation to proposers of reference material project; Part B: Guidelines and requirements for the implementation of reference material projects; Part C: Instructions for the preparation of BCR certifcation reports; EU, Standards, Measurements and Testing Programme, DOC 8CR/01/97, 15 April 1997

4.30. Quevauviller P and Maier E A (compiled); Kramer K J M (edited 1998) Production of Certified Reference Materials for Pollutants in Environmental Matrices. EC-Standards, Measurements and Testing Programme, Report Nr. EUR 18157 EN pp. 1–52

4.31. Guidelines for the Production and Certification of BAM Reference Materials (Part A, B, C), BAM, Berlin (21 March 1997)

4.32. ISO/REMCO N 330 List of producers of certifed reference materials, information by Task Group 3 "Promotion"

4.33. Wandelburg K (1987) Amts- und Mitteilungsblatt der Bundesanstalt für Materialforschung und -prüfung (BAM) 17, 47 1

4.34. Schimmel H (1998) GIT Laborfachzeitschrift **42**, 381

4.35. Recknagel S and Meier K (1997) GlT Labor-Fachzeitschrift **41**, 1164

4.36. Hospital P and Mittelstädt H (1996) in Tagungsband 21. Spektrometertagung, p. 291

4.37. DIN EN 10020 (1988), Beuth Verlag, Berlin

4.38. Tönsdorf H K , Kaestner W and Schnadt R (1989) Stahl und Eisen **109**, 743

4.39. Allen T (1997) Particle Size Measurement, 5th ed. Vol. **1**: Powder Sampling and Particle Size Measurement; Vol. **2**: Surface Area and Pore Size Determination, Chapman and Hall, London

4.40. Webb A and Orr C (1997) Analytical Methods in Fine Particle Technology, Micromeritics Instr. Corp., Norcross GA USA

4.41. Lowell S and Shields E (1991) Powder Surface Area and Porosity, 3rd ed., Chapman and Hall, London

4.42. Mikhail R Sh and Robens E (1983) Microstructure and thermal analysis of solid surfaces, John Wiley and Sons, Chichester

4.43. Schindler U and Wilson R (1980) Particulate Reference Materials. Production and Use of Reference Materials, Proc. Intern. Symp. held at the Bundesanstalt für Materialforschung und -prüfung (BAM), November 13–16, 1979

4.44. Müller R H and Schumann R (1996) Teilchengrößenmessung in der Laborpraxis, Wissenschaftliche Verlagsges. Stuttgart

4.45. Bernhardt C (1990) Granulometrie, Deutscher Verlag für Grundstoffindustrie, Leipzig

4.46. ISO DIS 9276-1 (1998) Representation of results of particle size analysis – Part 1: Graphical representation

4.47. ISO DIS 9276-2 (1997) Representation of results of particle size analysis – Calculation of average particle sizes/diameters and moments from particle size distributions

4.48. ISO DIS 9276-4 (1998) Representation of results of particle size analysis – Part 4: Characterization of a classification process used for particle size analysis

4.49. ISO DIS 13320 (1996) (E) Particle size analysis laser diffraction methods – Part 1: General principles

4.50. ISO DIS 133 17-1 (1998) Determination of particle size distribution by gravitational liquid sedimentation methods – Part 1: General principles and guidelines

4.51. Pankewitz A, Geers H and Röthele S (1998) Labor Praxis: **22**, 28

4.52. Mc Enaney B, Mays T J, Rouquerol J et al. (Eds.) (1997) Characterisation of porous solids IV (COPS IV) Proc. 4th IUPAC Symp. on Characterisation of Porous Solids, Bath, UK 15–18 September 1996 MPG Books Ltd, Bodmin, Cornwall

4.53. Rouquerol J et al (Ed.) (1994) COPS III: Marseille, France, 9–12 May 1993, Elsevier, Amsterdam

4.54. Rodriguez-Reinoso F et al (Ed.) (1991) COPS II: Alicante, Spain, 6–9 May 1990, Elsevier, Amsterdam

4.55. Unger K K et al (1988) COPS I: Bad Soden, Germany, 26–29 September 1987, Elsevier, Amsterdam

4.56. Sing K S W, Everett D H (1985) Pure & Appl Chem **57**, 603

4.57. DIN 66 131 (1993) Bestimmung der spezifischen Oberfläche von Feststoffen durch Gasadsorption nach Brunauer, Emmett und Teller (BET)

4.58. DIN 66 132 (1975) Bestimmung der spezifschen Oberfläche von Feststoffen durch Stickstoffadsorption; Einpunkt-Differenzverfahren nach Haul und Dümbgen

4.59. DIN 66 133 (1993) Bestimmung der Porenvolumenverteilung und der spezifischen Oberfläche von Feststoffen durch Quecksilberintrusion

4.60. DIN 66 134 (1997) Bestimmung der Porengrößenverteilung und der spezifischen Oberfläche mesoporöser Feststoffe durch Stickstoffsorption; Verfahren nach Barrett, Joyner und Halenda (BJH)

4.61. E DIN 66 135 (1998) Mikroporenanalyse mittels Gasadsorption: 66 135-1: Teil 1: Grundlagen und Meßverfahren; 66 135-2: Teil 2: Auswerteverfahren - Isothermenvergleichsverfahren; 66 135-3: Teil 3: Bestimmung des Mikroporenvolumens nach Dubinin und Radushkevich; 66 135-4: Teil 4: Bestimmung der Porenverteilung nach Horvath- Kawazoe und Saito-Foley

4.62. ASTM D 4404 (1992) Soil and rock by mercury intrusion porosimetry

4.63. Röhl-Kuhn B, Meyer K, Klobes P and Fritz T (1998) Fresenius J. Anal. Chem. **360**, 393

4.64. Bengtson A (1994) Spectrochimica Acta, **49B**, No. 4, pp. 411–429

4.65. ISO/TC201/WG1 study group paper: "Criteria for a Reference Material to Be Useful in GDOES Depth Profiling" by NIST/BAM/BHP/LECO

4.66. Winchester M R, Beck U Surface Interface Analysis (in press)

5 Reference Materials in Environmental Studies

Irene Nehls and Tin Win

Over the last few decades the character of chemical analytical investigations has changed significantly. Due to automatization and computerization in chemical measuring technology, an extensive amount of physical and chemical analysis data has been obtained in a relatively short time. These data have to be evaluated for soundness by means of statistical procedures because the results obtained may be used for decisions with weighty consequences.

High demands are made on environmental analysis regarding rightness, reliability and comparability of results, for the implementation of regulations concerning limiting values of pollutants and planning of remedial actions. Such demands require analytical laboratories to establish quality management systems.

Day to day practice has shown that the consequent use of validated and standardized analytical procedures and the strict implementation of systems of quality assurance is not sufficient to prevent laboratories producing incorrect and uncomparable analytical results.

What is the reason? The specific problem in environmental analysis is the complexity of the matrix and the variety of chemical compounds to be analysed, each bound differently into the matrix, subject to all the dependencies and influences of living nature.

Besides sampling on site, sample preparation and sample pretreatement depending on the specific matrix are the most important steps for ensuring correct analytical results.

A very significant tool for quality assurance in the analytical laboratory is the use of certified reference materials (CRM) and especially of certified matrix reference materials in the area of environmental studies.

The objective of this chapter is to describe the feasibility study and limits of the preparation and certification of matrix reference materials for organic environmental analysis.

5.1 Reference Materials in Environmental Analysis

Environmental analysis involves a wide spectrum of different kinds of reference materials. For the identification and characterization (qualitative and quantitative analysis) of organic and inorganic compounds in different environmental matrices, pure and ultrapure substances are used as calibration standards. The degree of purity is very important, for example, in trace analysis. These materials are employed mainly for spectrometric and chromatographic data (library) and for calibration of measuring instruments. Their production requires complicated syntheses involving many steps, followed by different clean-up procedures. After preparation of the substances, specifications for identification and purity must be fulfilled [5.1]. For identification, mass spectrometry (MS), nuclear magnetic resonance (NMR), infrared spectroscopy (IR), melting and boiling point, element analysis and refractive index methods are used. Purity is determined using methods such as gas chromatography, high performance liquid chromatography and differential scanning calorimetry (DSC). To determine stability of these pure standards, an extensive testing program is carried out to find the effects of temperature, light and air.

These pure substances, which are also available commercially, are the basis for preparation of calibration standards, dissolved in appropriate solvents at several concentration levels, in mixtures or in single form. Calibration of the measuring instruments serves for quantitative determination of compound contents in different environmental matrices as well as qualitative identification of the analytes. The quality of these calibration materials is very important. Calibration standards with certified purity are becoming more and more of importance in different catalogues. Most of the commercially available calibration standard solutions have no certified values. Systematic investigations have shown that the use of different quality levels of calibration standards considerably affects comparability of analytical results.

A second category, which is especially important for environmental analysis, is matrix reference materials. What is meant by the term matrix reference material? A matrix reference material is a material, which consists of a matrix similar to that found in nature or a real life matrix sample taken from a natural compartment and one or more analytes with certified values. This material must be sufficiently homogeneous and the matrix well enough characterized to ensure that the property values of the analyte may be traceable to the base unit SI or to a method described in national or international standards (i.e. method dependent). These materials can be used for the validation of analytical methods and also for quality assurance.

The above definition is derived from the ISO Guide 30 [5.2]. The environmental matrix has a decisive influence on sample preparation (grinding, sieving), the choice of extraction method (soxhlet, ultrasonic, microwave or accelerated solvent extraction), the clean-up procedure and the choice of the appropriate analytical method. Matrix reference materials thus fulfill the task of testing the rightness of the analytical procedure and all the steps it involves. These materials should not be used for calibration of the measuring instrument but only to check conformity of the specific analytical procedure used in the laboratory (in-house method).

Consumers usually ask national institutions to supply specific reference materials for their own purposes. Examples of such national bodies are the National Institute of Standards and Technology (NIST) in the USA, the Laboratory of the Government Chemist (LGC) in Great Britain, the Laboratoire National d'Essais (LNE) in France, the Nederlands Meetinstituut (NMi) in the Netherlands, the Federal Institute for Materials Research and Testing (BAM) in Germany, among others.

Two institutions of the Commission of the European Communities, the Community Bureau of Reference (BCR) and the Central Bureau for Nuclear Measurements (CBNM), fulfill a specific role relating to reference materials. The task of both these institutes is to supply Europe with reference materials, with an emphasis on socioeconomic relevance for the European community. The CBNM takes responsibility for reference materials in the nuclear and isotopic areas and the BCR for the environment, food, biology and technology. Especially in the environmental field, the BCR is the largest provider of matrix reference materials in Europe.

Over the last few decades most of BCR materials have been developed and produced within the scope of the Standards, Measurements and Testing (SM&T) programmes of the European Commission. These programmes were started in 1982 with a term of four years. At present, the Fifth Framework Programme (5FP) of the European Commission, started in 1999, is now underway.

Other institutes also offer reference materials for environment analysis. Examples are the Danish Center for Chemical Reference Materials (DANREF), the International Atomic Energy Agency (IAEA), and the Slovak Institute of Metrology (SIM).

In order to get a general idea of the certified reference materials available worldwide the reference materials service of the Laboratoire National d'Essais in Paris developed a computerised index called COde of Reference MAterials (COMAR) [5.3] in 1970. In collaboration with other European countries and also with the USA, Japan and China a database was compiled which now contains information on more than 10 000 reference materials from 20 countries (see also Chap. 7).

A current investigation of this recently updated database, searching for all available reference materials in the environmental area, reveals that from a total number of 1283 entries only 282 are for matrix reference materials, i.e. 22% (see Table 5.4).

Figure 5.1 shows the distribution of the 282 matrix reference materials in the areas organic, inorganic and other. Inorganic materials occupy the largest portion with 186 entries (66%). In contrast only 38 materials (14%) exist for the determination of organic substances in several environmental matrices, showing that these materials are lacking.

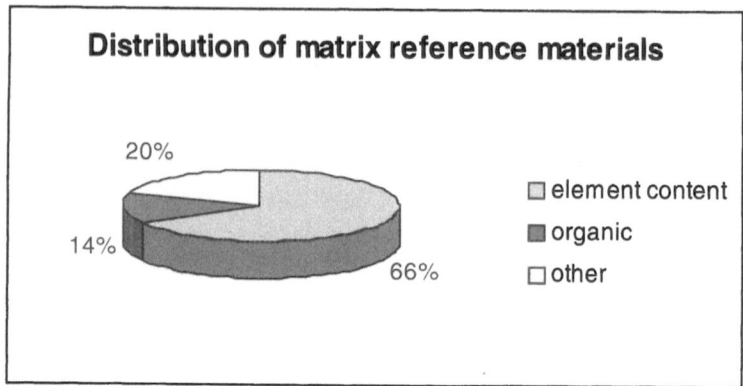

Fig. 5.1. Distribution of the matrix reference materials classified for certified properties

5.2   Matrix Reference Materials in Environmental Analysis

In the following, special features of the development, the procedure of preparation and certification of matrix reference materials are discussed. In all these steps stringent quality standards have to be respected, in conformity with European and international guidelines [5.4]. For the specific need of our institute "Guidelines for the production and certification of BAM reference materials" was adapted in 1997 [5.5]. These guidelines are the basis for the development and preparation of matrix reference materials in the area of organic and inorganic analysis.

Before the proper development of a special matrix reference material can begin, a few more practical questions have to be answered by every producer and user of reference materials:

- What will be the use of the reference materials?
- Analyte: which chemical species is to be determined qualitatively and quantitatively?
- Matrix: in which matrix is the analyte included? What is to be taken into account during the sample pretreatment and the following clean-up as far as the matrix is concerned?
- Performance characteristics: which requirements should be put on the measurement procedure to guarantee the correct level of rightness, precision, selectivity, linearity, specificity and robustness?
- Validation of methods: is there a routine analytical procedure for which reference materials are available? Must a new method be validated? Are new specific reference materials needed?
- Standard uncertainty: is the measurement uncertainty known or can it be estimated? Must we consider the uncertainty of the whole measurement procedure or only of a partial step?

Matrix reference materials for environmental analysis can be classified into two groups:

- spiked matrix reference materials;
- natural matrix reference materials.

The first approach is to spike a basic material with definite pure substances or mixtures of appropriate concentrations. Pure reference materials or standard substances are taken often as the basis for the preparation and verification of all spiked materials. The homogeneous distribution of these spikes within the basic materials is the main point of this procedure. The Environmental Institute of the Joint Research Center in Ispra (Italy) has extensive experience in this area [5.6].

Strong requirements must be set on matrix reference materials, such as homogeneity, stability and similarity with the real life sample. The last requirement in particular is difficult to fulfill in the case of spiked matrix reference materials. The problem is to obtain binding characteristics similar to those of naturally aged real samples.

This situation becomes much better when natural materials are taken in to account. Homogeneity and stability are the decisive properties for candidate reference materials. Similar to the case of spiked reference materials, they have to be tested for homogeneity at two different stages of production, in the bulk material and after packaging (bottling or ampouling). Homogeneity must be tested both within individual bottles and from one bottle to another.

In our group we only deal with the preparation and certification of *natural* matrix reference materials for environmental analysis and especially for organic compounds.

5.2.1 Preparation Procedure for Matrix Reference Materials

In accordance with the "Guidelines for the production and certification of BAM reference materials" a scheme of work programme has to be set up. The most important steps in this scheme are the selection of the natural material, drying and sieving, homogenization, packaging and determination of the stability of the chemical parameters.

The first step is the *selection* of a suitable natural material. According to the requirements of environmental monitoring programmes it must be decided which analyte in which matrix is needed for validation of methods or for quality assurance. Based on numerous experiments it was decided to take only "naturally contaminated" samples in the case of solid materials. Investigations have shown that there are considerable differences in the extraction behaviour and in the recovery rate. Attempts at artificial ageing are not yet completed.

Furthermore it must be tested whether the analyte concentrations to be determined can be found in daily practice and which problems can arise with the matrix specific extraction.

The most important question is that of homogeneity. A reference material can only be prepared if the selected material can be homogenized.

Particularly in the area of organic materials, the stability of analytes as well as that of the matrix itself, plays a decisive role in the process of development and certification of a reference material. Materials have to be observed at different temperatures and over a long period.

After the determination of all available chemical and physical parameters for the characterization of materials taken from the environmental compartment, the candidate material is selected. The choice of candidate reference materials depends on a number of characteristics, such as particle size and morphology, moisture content, organic carbon content, particle density and the concentration of the analytes in the different particle fractions.

The next step is a careful *drying* in air or by means of freeze drying. All the first analytical data refer to the bulk material with a particle size of < 2 mm.

Separation into different particle fractions is carried out by means of *sieving*. In accordance with our strategy the materials are sieved in narrow particle fractions of 2–1 mm, 1–0.5 mm, 0.5–0.25 mm, 0.25–0.125 mm and 0.125–0.063 mm. Within each of these five fractions, the level of contamination is determined individually.

In the case of soils, the particle fractions in the range from 0.5 to 0.063 mm are the most frequent in nature, independently of the type of soil, as geologists will confirm. The middle fractions of particle size of 0.5–0.25 mm and 0.25–0.125 mm are chosen as preferred candidate materials.

After sieving the statistical *homogenization* of each batch is carried out according to a four step scheme using a rotational sample divider, the cross-riffling method [5.7]. The scheme works with an eightfold sample division so that sample numbers of 128, 256 or 512 in each batch can be realized.

The principle of the materials division can be illustrated by the following example. Starting with 20.5 kg raw material, the division steps lead to the following results:

1. division step   ⟶ 8 × 2.56 kg (partial sample 01 ...   08),
2. division step   ⟶ 64 × 320 g (partial sample 0101 ...   0808),
3. cross-riffling step ⟶ 8 × 2.56 kg (partial sample A ...   H),
4. division step   ⟶ 64 × 320 g (partial sample A.01 ...   H.08),
5. division step   ⟶ 512 × 40 g (partial sample A.0101 ... H.0808).

A larger number of samples is possible but we are not convinced that it should be the aim of our work, for several reasons. The first is the limited stability of the organic contaminants in different environmental matrices. Sensitive materials have to be stored at −20°C in a freezer. This storage requires a great deal of energy and causes high costs, and hence high prices for the materials. Furthermore, stability of the samples must be regularly checked. In cases where changes of concentration are found, a new represen-

Table 5.1. Scheme of sample division corresponding to the cross-riffling method [5.7]

01	02	03	04	05	06	07	08	
0101	0202	0303	0404	0505	0606	0707	0808	A
0102	0203	0304	0405	0506	0607	0708	0801	B
0103	0204	0305	0406	0507	0608	0701	0802	C
0104	0205	0306	0407	0508	0601	0702	0803	D
0105	0206	0307	0408	0501	0602	0703	0804	E
0106	0207	0308	0401	0502	0603	0704	0805	F
0107	0208	0301	0402	0503	0604	0705	0806	G
0108	0201	0302	0403	0504	0605	0706	0807	H

tative analysis has to be done for the whole batch. If the sample is already certified then recertification has to be carried out.

Our experience shows that the preparation and also the certification of smaller sample numbers has the advantage that after two or three years all materials are used up. It seems better to develop and prepare diverse materials in smaller numbers with different matrices and contamination content. In this way it is possible to be more flexible in conjunction with the needs of laboratories working in environmental analysis. Over the last few decades, various new organic contaminants have been detected in the environment, harmful and dangerous for both human beings and animals. The user of reference materials expects to obtain a wide spectrum of different materials from the producer, and this is the biggest problem to be solved and an exceptional challenge.

After statistical homogenization, the material is put in brown glass bottles and stored in the freezer at $-20°C$. The next step is *determination of the homogeneity*. Approximately 10% of the samples of a batch are taken for the measurements, using the BAM programme HOMEX for data evaluation.

The *determination of the stability* of the chemical parameters is performed at different temperatures. We choose five different temperature levels for the stability investigation: $+70$, $+40$, $+20$ (ambient temperature), $+4$ and $-20°C$. Depending on the selected temperature the investigation extends over a period of 8–10 months at $70°C$, one year at $+40°C$, two years at $+20°C$ and $+4°C$, whilst at $-20°C$ the materials should be stable over a minimum period three years.

5.2.2 Certification of Matrix Reference Materials

The certification procedure itself is the last and also the most complicated step in the process of development and preparation of reference materials.

Corresponding to the "Guidelines for the production and certification of BAM reference materials" [5.5], the certification procedure can follow different approaches.

It was decided to carry out the certification procedure by interlaboratory study with qualified laboratories using one or more methods of demonstrated accuracy. A feasibility study was carried out with an orientation interlaboratory comparison to determine the qualified laboratories. After that, the certification interlaboratory comparison could begin.

In the following two examples, the steps of the certification procedure are shown. The first example is a matrix reference material (BAM CRM 2002) for the determination of mineral oil content in soil. In the second example, a reference material (BAM CRM 5001) is described which can be used for the determination of polychlorinated biphenyls in waste oil. The certified values are traceable to a method described in the international standard.

BAM CRM 2002

Origin, Preparation and Description of the Reference Material. BAM CRM 2002 is a natural matrix reference material taken from a diesel-fuel contaminated site in South Germany. CRM 2002 is supplied in brown glass bottles containing approximately 28 g of material. The bottles are screw-capped and further tightened with PTFE foils. Homogeneity tests revealed an inhomogeneity, corresponding to a relative standard deviation not exceeding 3%. This inhomogeneity has been taken into account for the calculation of the combined uncertainty. The tests for homogeneity and stability are described in detail in the certification report.

Intended Use. The reference material CRM 2002 may be used to assess the implementation of the analytical method described in ISO/TR 11046 (part A) [5.8] or to verify the performance of method modifications.

The material has to be stored at $-20°$C. Before withdrawing small amounts/portions of the material, the bottle is warmed to room temperature and the contents well mixed before sampling. After sampling, the lid is firmly screwed on (airtight) and it is stored at $-20°$C. Based on the stability tests carried out, a minimum material stability of 5 years may be expected. If future stability checks should reveal significant changes in the certified properties, BAM will advise all known customers of the CRM under consideration and recertify the values as soon as possible. The recommended minimum sample amount for all kinds of applications is 2 g.

The analytical method used for certification corresponds to ISO/TR 11046 (part A) and implies extraction, clean-up and determination by IR spectroscopy. In the certification study, different extraction techniques have been used. Details of the method are given in the certification report.

Evaluation of Results. The certified value of 1750 mg/kg was obtained for CRM 2002 as the mean of means with a standard deviation of the mean of 50 mg/kg. The laboratory means (Xi) are based on 6 replicate measurements (n). Each, these individual values, the lab means and the corresponding standard deviations (Si) and relative standard deviations (Si, rel.) are given in Table 5.2.

Table 5.2. Summary of the laboratory means and individual values

Lab nr.	Mean Xi [mg/kg]	Si [mg/kg]	Si, rel. [%]	n	Individual values 1	2	3	4	5	6
1	1472.83	43.06	2.92	6	1469	1472	1433	1556	1455	1452
2	1587.83	46.25	2.91	6	1522	1639	1541	1620	1609	1596
3	1611.00	90.47	5.62	6	1730	1571	1568	1560	1517	1720
4	1671.50	68.12	4.08	6	1618	1624	1626	1648	1734	1779
5	1700.67	6.28	0.37	6	1704	1705	1689	1703	1698	1705
6	1718.33	16.13	0.94	6	1705	1716	1724	1706	1711	1748
7	1728.33	91.09	5.27	6	1640	1870	1810	1660	1700	1690
8	1747.83	44.73	2.56	6	1711	1740	1705	1727	1817	1787
9	1900.33	34.22	1.80	6	1881	1866	1881	1887	1940	1947
10	1906.67	42.60	2.23	6	1883	1884	1907	1914	1866	1986
11	1988.83	11.62	0.58	6	1989	1981	2006	1998	1985	1974
12	2015.50	32.01	1.59	6	2047	2029	1980	2045	2019	1973

The lab means, the certified value and the corresponding confidence intervals (at the 95% confidence level) are plotted in the diagram below (Fig. 5.3).

According to [5.9], the effective number of degrees of freedom for the combined uncertainty of the certified value was established to be 26 using the expression of Welch and Satterthwaite (with 15 degrees of freedom for the inhomogeneity contribution to the combined uncertainty, c.f. [5.10]). Therefore, $t_{26,0.05}$ is 2.056, and the 95% confidence interval has a half-width of 150 mg/kg.

The evaluation of certification study results and the certification of CRM 2002 have been carried out in full accordance with the "BCR Guidelines for the production and certification of reference materials" [5.4] and are described in detail in the certification report.

Note. A detailed report of the production of the samples, the analytical procedures applied and the evaluation of the analytical data can be requested from BAM by the customer. Non-certified parameters characterizing secondary properties of the soil are given in Sect. 5.3 of the report.

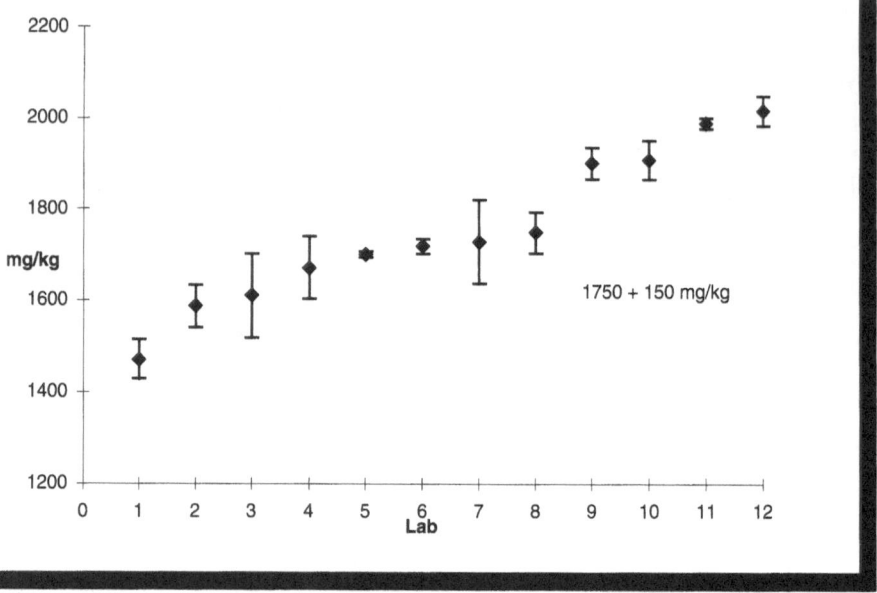

Fig. 5.2. Results of lab means, the certified value and the corresponding confidence intervals (at the 95% confidence level)

BAM CRM 5001

Origin, Preparation and Description of the Reference Material.
BAM CRM 5001 is a PCB-containing waste oil taken from a transformer. It is colourless, transparent and without any suspended particles. The material is available as a kit together with the blank transformer oil BAM CRM 5001/blancoil to be used for matrix simulation. The reference material is supplied in brown glass ampoules containing approximately 15 g oil. The material is homogeneous. According to the stability tests, a minimum material stability of five years can be expected. If periodic stability checks should show significant changes in the certified properties, BAM will inform all known customers of the CRM and replace the material with new certified units as soon as possible. The tests for homogeneity and stability are described in detail in the certification report.

Intended Use. The reference material BAM CRM 5001 in combination with the BAM CRM 5001/blancoil can be used to assure the correct implementation of the analytical method described in CEN draft proposal "Petroleum products and used oils – Determination of polychlorinated biphenyls" from 01.11.1993 [5.11].

The contents of eight PCB congeners are certified. The separation of PCB 101 from 84 and 138 from 163 could not be done with the specified GC columns, and sums are therefore given. The content of PCB congeners 18, 28, 31 and 194 are below the detection limit.

The material has to be stored at 5°C. Normal laboratory safety precautions should be taken when opening the glass ampoule. After opening, the contents should be transferred to a clean glass container with a ground glass stopper. The recommended minimum sample amount for all kinds of application is 1 g.

The analytical method used for certification corresponds to CEN draft proposal "Petroleum products and used oils – Determination of polychlorinated biphenyls" [5.11]. This draft proposal specifies a method for determining the content of 12 PCB congeners (PCB 18, 28, 31, 44, 52, 101, 118, 138, 149, 153, 180 and 194). For matrix simulation, a PCB-free oil has been used.

In accordance with this method, solutions of PCB-free oil spiked with PCB standard solutions should be treated in exactly the same way as the sample. The PCB congeners are determined by gas chromatography using a high efficiency narrow-bore capillary column, an electron capture detector and an internal standard.

Evaluation of Results. Certification and evaluation of certification study results were carried out according to the guidelines [5.5]. The statistical analysis was executed with the BCR computer programme [5.12].

The certified values of the contents of eight PCB congeners are the mean values of means. The laboratory means result from six replicate measurements each; these lab means and the corresponding statistical deviations are given in Table 5.3.

In the certification interlaboratory comparison it was not possible for the laboratories to separate the overlapping PCB 84/101 and PCB 138/163, respectively, with the gas chromatographic columns used here. For these PCB congeners, sums are given in the table.

Table 5.3. Mean of means (Xi) and standard deviations (Si and Si, rel.) for PCB congeners

PCB (IUPAC nr.)	Lab nr.	1	2	3	4	5	6	7	8	9
44	Mean [μg/kg]	234.50	237.00		187.00	190.83	381.33	283.33	203.50	195.17
	Si [μg/kg]	3.67	15.95		6.60	25.10	124.93	19.45	10.01	2.79
	Si, rel. [%]	1.57	6.73		3.53	13.15	32.76	6.86	4.92	1.43
52	Mean [μg/kg]	769.00	951.33	653.50		829.00	545.67	895.00	742.17	919.00
	Si [μg/kg]	38.68	20.05	167.25		15.50	33.77	70.99	76.63	22.63
	Si, rel. [%]	5.03	2.11	25.59		1.87	6.19	7.93	10.32	2.46
101+84	Mean [μg/kg]	1237.6	1579.3	1733.5	1173.5	1073.0	1557.5	1391.6	1569.6	1519.8
	Si [μg/kg]	71.39	41.68	56.77	39.69	28.64	37.58	46.22	30.32	17.08
	Si, rel. [%]	5.77	2.64	3.28	3.38	2.67	2.41	3.32	1.93	1.12
118	Mean [μg/kg]	780.83	1054.6	1027.6	721.50	794.83	852.67	944.17	751.33	837.50
	Si [μg/kg]	53.02	32.40	31.79	18.94	15.52	54.08	38.45	17.53	12.57
	Si, rel. [%]	6.79	3.07	3.09	2.63	1.95	6.34	4.07	2.33	1.50
138+163	Mean [μg/kg]	828.33	899.50	812.67	735.50	768.50	741.67	854.33	805.83	755.67
	Si [μg/kg]	57.03	41.06	34.19	20.05	16.88	53.82	27.24	11.43	7.09
	Si, rel. [%]	6.88	4.56	4.21	2.73	2.20	7.26	3.19	1.42	0.94
149	Mean [μg/kg]	692.33	809.67	732.83	595.33	531.50	610.83	626.83	542.17	685.33
	Si [μg/kg]	43.33	21.67	23.16	11.78	16.21	14.50	31.05	14.55	15.73
	Si, rel. [%]	6.26	2.68	3.16	1.98	3.05	2.37	4.95	2.68	2.30
153	Mean [μg/kg]	695.83	723.33		692.33	616.67	772.00	638.67	672.67	761.00
	Si [μg/kg]	51.89	27.84		19.26	8.64	67.08	28.65	13.13	9.47
	Si, rel. [%]	7.46	3.85		2.78	1.40	8.69	4.49	1.95	1.24
180	Mean [μg/kg]	97.00	101.83	133.83	104.67	98.67	130.50	143.67	80.83	
	Si [μg/kg]	7.21	5.91	9.06	1.03	5.99	15.14	3.39	2.48	
	Si, rel. [%]	7.43	5.81	6.77	0.99	6.07	11.60	2.36	3.07	

5.3   Use of Matrix Reference Materials

Only a small part of all matrix reference materials developed in our group are certified and commercialized. Most matrix reference materials are used for interlaboratory comparisons and for the validation of new analytical methods.

5.3.1   Use of Certified Matrix Reference Materials

Certified matrix reference materials are used exclusively for internal quality assurance. Laboratories have the possibility of checking the rightness and reliability of their own results. The whole analytical procedure, starting from sample pretreatment, especially extraction and clean-up as well as measurement, can be tested.

These certified materials are intended to be used when a laboratory is installing this specific method, to monitor the testing capability of the laboratory over time or when new staff members use the method for the first time.

5.3.2   Reference Materials for Validation of Analytical Methods

Thoroughly homogeneous and well characterized samples, i.e. reference materials, are to be used in interlaboratory comparisons for the determination of performance characteristics such as reproducibility, comparability, robustness and measurement uncertainty of a method which is to be standardized (DIN, CEN, ISO, etc.).

Reference materials with certified reference values traceable to the amount of the substance on the basis of SI units can be used to validate the in-house method or a newly developed method of a laboratory.

5.3.3   Reference Materials for Proficiency Testing

Reference materials can be used for the proficiency testing of laboratories in the accreditation procedure, to assess their technical competence in carrying out a specific test method. According to the definition of ISO/IEC Guide 43 [5.13], proficiency testing is the determination of a laboratory's testing performance by means of interlaboratory comparisons.

Interlaboratory comparison is defined as "organisation, performance and evaluation of tests on the same or similar test items by two or more laboratories in accordance with predetermined conditions". The requirement set for the test items concerning homogeneity and short term stability is the same for the reference materials, so that in some proficiency testing programmes these reference materials have been used.

For the assessment of the technical competence of a laboratory in an interlaboratory comparison, a criteria for the performance score is set which should be fit for the purpose, and against which the degree of the deviation of the laboratory results from the assigned values can be judged.

There are various procedures available for establishing assigned values. In most of the proficiency testing programmes with naturally contaminated samples, consensus values using statistics described for example in ISO 5725-2 [5.14], or robust statistical techniques, are taken as assigned values. In some cases, as in the case of the International Measurement Evaluation Programmes (IMEP) organized by the Joint Research Center (IRMM) in Geel (Belgium), the certified reference materials were used. The participating laboratories recieve a certified test sample (with undisclosed concentration values) to be analysed by the laboratory using its own routine analytical procedure. The measurement results of the participants are evaluated against the metrological reference values traceable to SI systems. The quality of the results of the laboratories can be assessed in comparison with IMEP reference values.

Bilateral Proficiency Testing. In accreditation procedures, where an appropriate proficiency testing programme is not available to evaluate a laboratory's testing performance in a specific field, the laboratory recieves a test item with precisely determined characteristics, i.e. a reference material with certified values (values undisclosed to the laboratory) from the assessor or provided by a third party. The results of the laboratory are then evaluated against the reference values with a performance criteria set by the accreditation body/ organising body which is fit for the purpose.

Table 5.4. COMAR 1998. Summary of registered matrix reference materials in the category "Reference materials for quality of life" inclusive of environment and agriculture. A list of producers is given in Table 5.5

Matrix material	Producer	Content	Code	Year
Fly ash from pulverised coal	BCR	Element content	CRM 038	
Aquatic plant	BCR	Element content	CRM 060	1982
Aquatic moss	BCR	Element content	CRM 061	1982
Olive leaves	BCR	Element content	CRM 062	1982
Calcareous loam soil	BCR	Element content	CRM 141	1983
Light sandy soil	BCR	Element content	CRM 142R	1994
Sewage sludge amended soil	BCR	Element content	CRM 143R	1994
City waste incineration ash	BCR	Element content	CRM 176	1984
Cod liver oil	BCR	PCB (7)	CRM 349	1987
Mackerel oil	BCR	PCB (7)	CRM 350	1987
Sea lettuce	BCR	Element content	CRM 279	1987
Mussel tissue	BCR	Element content	CRM 278	1988
Rye grass	BCR	Element content	CRM 281	1988
Estuarine sediment	BCR	Element content	CRM 277	1988
River sediment	BCR	Element content	CRM 320	1988
Beech leaves	BCR	Element content	CRM 100	1993
Spruce needles	BCR	Element content	CRM 101	1989
Seawater	BCR	Element content	CRM 403	1991
Freshwater	BCR	Element content	CRM 398	1991
Freshwater	BCR	Element content	CRM 399	1991
Plankton	BCR	Element content	CRM 414	1992
Waste mineral oil	BCR	PCB (10)	CRM 420	1992
Cod muscle	BCR	Element content	CRM 422	1992
Rainwater	BCR	Mineral (low)	CRM 408	1993
Rainwater	BCR	Mineral (high)	CRM 409	1993
Sewage sludge of mixed origin	BCR	S, element content	CRM 145R	1993
Waste mineral oil	BCR	PCB (9)	CRM 449	1993
Fly ash	BCR	Dioxins (5), furans (6)	CRM 429	1993
Dried sewage sludge	BCR	PAH (8)	CRM 088	1994
Fly ash on artificial filters	BCR	Element content	CRM 128	
Coal powder	BCR	Total fluorine	CRM 460	
Clay	BCR	Fluorine	CRM 461	
Coastal sediment	BCR	Tributyltin, dibutyltin	CRM 462	
Industrial harbour sediment	BCR	Tributyltin	CRM 424	1994
Tuna fish	BCR	Total and methylmercury	CRM 463	1994
Tuna fish	BCR	Total and methylmercury	CRM 464	1994
Industrial soil	BCR	Chlorobiphenyls (7)	CRM 481	1994
Estuarine water	BCR	Element content	CRM 505	1994

Table 5.4. (continued)

Matrix material	Producer	Content	Code	Year
Fresh water	BCR	Nitrate	CRM 479	1994
Fresh water	BCR	Nitrate	CRM 480	1994
Lichen	BCR	Element content	CRM 482	
Fly ash	BCR	Dioxins, furans	CRM 490	
Industrial soil	BCR	PAH (9)	CRM 524	
Aquatic plant	BCR	Element (Cr)	CRM 596	
Sewage sludge (industrial origin)	BCR	Element content	CRM 146R	
Sewage sludge	BCR	Element (Cr)	CRM 597	
Sewage sludge	BCR	Element content	CRM 145	1983
Sewage sludge (domestic origin)	BCR	Element content	CRM 144	1985
Sewage sludge (industrial origin)	BCR	Element content	CRM 146	1983
Lake sediment	BCR	Element content	CRM 280	1988
Cotonnier	INRA	Element content		1975
Codia	INRA	Element content		1975
Eucalyptus	INRA	Element content		1975
Foin	INRA	Element content		1985
Herea	INRA	Element content		1975
Mais	INRA	Element content		1975
Olivier	INRA	Element content		1975
Oranger	INRA	Element content		1975
Pecher	INRA	Element content		1975
Pommier golden	INRA	Element content		1975
Riz	INRA	Element content		1980
Salade	INRA	Element content		1985
Tabac	INRA	Element content		1980
Vigne	INRA	Element content		1975
Marine material	IAEA	Chlorinated hydrocarbons	MA-A-1/OC	
Marine material	IAEA	Element content	MA-A-1/TM	
Feldspar	IAEA	Elements (U, K)	F-1	
Lake sediment	IAEA	Element content	SL-1	
Lake sediment	IAEA	Radioactivity (K-40. Cs-137)	SL-2	1986
Mediterranean seaweeds	IAEA	Activity of radionuclides	IAEA-308	1988
Shrimp homogenate	IAEA	Organochlorines	MA-A-3/OC	
Stream sediment	IAEA	Radioactivity (Ra-226)	IAEA-313	1988
Stream sediment	IAEA	Radioactivity (Ra-226)	IAEA-314	1988
Soil	IAEA	Radioactivity (Ra-226)	IAEA-312	1988
Fish tissue lyophilised	IAEA	Element content	MA-B-3/TM	
Pacific ocean sediment	IAEA	Radionuclides	IAEA-367	1990
Pacific ocean sediment	IAEA	Radionuclides	IAEA-368	1990
Pacific ocean water	IAEA	Radionuclides	IAEA-298	

Table 5.4. (continued)

Matrix material	Producer	Content	Code	Year
Mussel homogenate	IAEA	Mercury, methylmercury, PAH	IAEA-142	1996
Polluted marine sediment	IAEA	Element content	IAEA-356	1995
Lichen	IAEA	Element content	IAEA-336	1996
Baltic sea sediment	IAEA	Radionuclides	IAEA-300	1993
Fish flesh homogenate	IAEA	Chlorinated hydrocarbons	MA-A-2/OC	
Fish flesh homogenate	IAEA	Chlorinated hydrocarbons	MA-A-2/TM	
Mussel homogenate	IAEA	Chlorinated hydrocarbons	MA-M-2/OC	
Marine algae	IAEA	Radionuclides	AG-B-1	1984
Sea plant	IAEA	Chlorinated hydrocarbons	SP-M-1/OC	
Marine sediment	IAEA	Radionuclides	SD-N-1/2	1985
Marine sediment	IAEA	Radionuclides	SD-N-2	1985
Marine sediment	IAEA	Chlorinated hydrocarbons	SD-M-1/OC	
Lake sediment	IAEA	Element content	SL-3	
Fresh water	IAEA	Element content	W-4	
Fish tissue (lyophilised)	IAEA	Chlorinated hydrocarbons	MA-B-3/OC	
Sea plant	IAEA	Radionuclides	IAEA-307	1988
Marine sediment (lyophilised)	IAEA	Element content	SD-M-2/TM	
Coastal sediment	IAEA	Organic compounds	IAEA-357	1991
Buffalo river sediment	NIST	Element content	SRM 2704	1990
Filter media	NIST	Element content	SRM 3087	1990
Filter media	NIST	Elements (Be, As)	SRM 2677a	1990
Marine sediment	NIST	Elements, PAH, PCB, pesticides	SRM 1941a	1994
Urban particulate matter	NIST	Element content	SRM 1648	1982
Estuarine sediment	NIST	Element content	SRM 1646a	1995
Simulated rainwater	NIST	Element content	SRM 2694	1986
Filter media	NIST	Element content	SRM 2676a	1987
Urban dust	NIST	PAH (5)	SRM 1649	1982
Residual fuel oil	NIST	Suphur	SRM 1621d	1991
River sediment	NIST	PCB	SRM 1939	1990
Tennessee river sediment	NIST	Mercury	RM 8406	1990
Tennessee river sediment	NIST	Mercury	RM 8407	1990
Tennessee river sediment	NIST	Mercury	RM 8408	1990
Pine needles	NIST	Element content	SRM 1575	1976
Tomato leaves	NIST	Element content	SRM 1573 a	1995
Apples leaves	NIST	Element content	SRM 1515	1991
Peach leaves	NIST	Element content	SRM 1547	1991
Citrus leaves	NIST	Element content	SRM 1572	1982
Water	NIST	Mercury	SRM 1641c	1993
Wheat flour	NIST	Element content	SRM 1567a	1988
Rice flour	NIST	Element content	SRM 1568a	1995

Table 5.4. (continued)

Matrix material	Producer	Content	Code	Year
Bovine liver	NIST	Element content	SRM 1577b	1991
Montana soil	NIST	Element content	SRM 2710	1993
Montana soil	NIST	Element content	SRM 2711	1993
San Joaquin soil	NIST	Element content	SRM 2709	1993
Spinach leaves	NIST	Element content	SRM 1570a	1994
Hard drinking water	LGC	Anions	LGC 6012	1998
Soft drinking water	LGC	Anions	LGC 6013	1997
Water	LGC	Triazines	LGC 1004	1995
Water	LGC	Urons	LGC 1005	1995
Petroleum products	LGC	Dibuthyl sulphide	LGC 4000	1994
Petroleum products	LGC	Dibenzothiophene	LGC 4001	1994
Soil	LGC	Element content	LGC 6135	1997
Coal	LGC	Element content, PAH	LGC 6138	1997
River clay sediment	LGC	Element content	LGC 6139	1997
Landfill leachate	LGC	Element content	LGC 6175	1997
Landfill leachate	LGC	Anions	LGC 6176	1997
Hard drinking water	LGC	Element content	LGC 6010	1998
Soft drinking water	LGC	Element content	LGC 6011	1998
Gas oil	LGC	Sulphur	LGC 3000	1995
Gas oil	LGC	Sulphur	LGC 3001	1994
Potato	LGC	Sulphur dioxide	LGC 7111	1997
Pond sediment	NIES	Element content	NIES No.2	1981
Vehicle exhaust particulates	NIES	Element content	NIES No.8	1987
Sargasso	NIES	Element content	NIES No.9	1988
Rice fluor-unpolished	NIES	Element content	NIES No.10a	1989
Rice fluor-unpolished	NIES	Element content	NIES No.10b	1989
Rice fluor-unpolished	NIES	Element content	NIES No.10c	1989
Fish tissue	NIES	Tributhyltin, triphenyltin	NIES No.11	1990
Chlorella	NIES	Element content	NIES No.3	1982
Tea leaves	NIES	Element content	NIES No. 7	1986
Water	NRC	Lead	GBW 08601	1987
Water	NRC	Cadmiun	GBW 08602	1987
Water	NRC	Mercury	GBW 08603	1987
Water	NRC	Fluoride	GBW 08604	1987
Water	NRC	Arsenic	GBW 08605	1987
Water	NRC	Metal element	GBW 08607	1988
Water	NRC	Metal element	GBW 08608	1988
Coal fly ash	NRC	Fluoride	GBW 08402	1991
Corn	NRC	Fluoride	GBW 08506	1991
Corn	NRC	Fluoride	GBW 08507	1991
Simulated rainwater	NRC	pH, conductivity	GBW 08627	1995

Table 5.4. (continued)

Matrix material	Producer	Content	Code	Year
Simulated rainwater	NRC	pH, conductivity	GBW 08628	1995
Simulated rainwater	NRC	pH, conductivity	GBW 08629	1995
Rice	NRC	Cadmium	GBW 08510	1997
Rice	NRC	Cadmium	GBW 08511	1997
Rice	NRC	Cadmium	GBW 08512	1997
Wheat flour	COCRI	Element content	GBW 08503	1987
Cabbage	FDSI	Element content	GBW 08504	1987
Pork liver	FDSI	Element content	GBW 08551	1987
Rice	FDSI	Mercury	GBW 08508	1991
Prawn	FDSI	Element content	GBW 08572	1992
River sediment	IEC	Element content	GBW 08301	1986
Xizang soil	IEC	Element content	GBW 08302	1987
Coal fly ash	IEC	Element content	GBW 08401	1986
Peach leaves	IEC	Element content	GBW 08501	1986
Tea	IEC	Element content	GBW 08505	1989
Mussel	IEC	Element content	GBW 08571	1991
Coal fly ash	IEC	PAH (5)	GBW 08403	1996
Pork muscle	SINR	Element content	GBW 08552	1995
Bush branches and leaves	IGGE	Element content	GBW 07602	1991
Bush branches and leaves	IGGE	Element content	GBW 07603	1991
Poplar leaves	IGGE	Element content	GBW 07604	1991
Tea	IGGE	Element content	GBW 07605	1991
Water	SITT	Mercury	GBW 08609	1988
Polluted farm land	BMEMC	Element content	GBW 08303	1988
Rice flour	BMEMC	Element content	GBW 08502	1986
Offshore marine sediment	SMI	Element content	GBW 07314	1994
Soil	EMCHJ	Element content	GBW 07409	1993
Soil	EMCHJ	Element content	GBW 07410	1993
Soil	EMCHJ	Element content	GBW 07411	1993
Sludge from city water treatement	PB-ANAL	Element content	SIRM 12-3-12	1996
Sludge from city water treatement	PB-ANAL	Element content	SIRM 12-3-13	1996
Sludge from city water treatement	PB-ANAL	Element content	SIRM 12-3-14	1996
Brown coal fly ash	PB-ANAL	Element content	SIRM 12-1-01	1995
Fly ash EOP	PB-ANAL	Element content	SIRM 12-1-02	1985
Fly ash ECH	PB-ANAL	Element content	SIRM 12-1-03	1985
Fly ash ECO	PB-ANAL	Element content	SIRM 12-1-04	1985
Steel plant flue dust	PB-ANAL	Element content	SIRM 12-1-05	1988
Copper plant flue dust	PB-ANAL	Element content	SIRM 12-1-06	1989
Green algae	PB-ANAL	Element content	SIRM 12-2-02	1991
Lucerne	PB-ANAL	Element content	SIRM 12-2-03	1991
Fly ash	PB-ANAL	Radioactivity	SIRM 11-2-01	1990

Table 5.4. (continued)

Matrix material	Producer	Content	Code	Year
Wheat bread flour	PB-ANAL	Element content	SIRM 12-2-04	1992
Rye bread flour	PB-ANAL	Element content	SIRM 12-2-05	1992
Soil eutric cambisols	PB-ANAL	Element content	SIRM 12-1-07	1992
Soil orthic luvisols	PB-ANAL	Element content	SIRM 12-1-08	1992
Soil rendzina	PB-ANAL	Element content	SIRM 12-1-09	1992
Paper	PPRI	Element content	SIRM 12-2-12	1995
Pulp	PPRI	Element content	SIRM-12-2-13	1995
Water	RII	Element content	SIRM 12-3-10	1996
Carnation dianthus	WAU	Element content	B 112	1992
White cabbage	WAU	Element content	B 111	1992
Maize (plant)	WAU	Element content	B 211	1992
Gladiolus (leaf)	WAU	Element content	B 212	1992
Conifers	WAU	Element content	B 213	1992
Wheat (straw)	WAU	Element content	B 214	1992
Cabbage (leaf)	WAU	Element content	B 215	1992
Broad beans	WAU	Element content	B 216	1992
Fennel (turnip)	WAU	Element content	B 217	1992
Tabacco (leaf)	WAU	Element content	B 218	1992
Banana (fruit)	WAU	Element content	B 219	1992
Crocus	WAU	Element content	B 220	1992
Gherkin (mixture)	WAU	Element content	B 221	1992
Apple (leaf mixture)	WAU	Element content	B 222	1992
Grass (mixture)	WAU	Element content	B 223	1992
Seaclub-rush	WAU	Element content	B 224	1992
Cord grass	WAU	Element content	B 225	1992
Sea aster	WAU	Element content	B 226	1992
Wheat (straw)	WAU	Element content	B 227	1992
Tall fescue	WAU	Element content	B 228	1992
Lucerne	WAU	Element content	B 212	1992
Maize (stalk)	WAU	Element content	B 230	1992
Pine (needles)	WAU	Element content	B 231	1995
Potato	WAU	Element content	B 232	1995
Winter carrots	WAU	Element content	B 233	1995
Endive mixture	WAU	Element content	B 234	1995
Curly kale (leaf)	WAU	Element content	B 235	1995
Scots pine (needles)	WAU	Element content	B 236	1995
Sewage sludge	RIZA	Element content	CRMPR 9472	1994
Soil	RIZA	Total petroleum hydrocarbons	PR 9583	1995
Soil	RIZA	BTEX	PR 9584	1995
Polluted sediment	RIZA	Element content	PR 96962	1996
Non-polluted sediment	RIZA	Element content	PR 96961	1996

Table 5.4. (continued)

Matrix material	Producer	Content	Code	Year
Spruce twigs	CCRMP	Element content	CLV-1	1985
Spruce needles	CCRMP	Element content	CLV-2	1985
Lake sediment	CCRMP	Element content	LKSD-1	1989
Lake sediment	CCRMP	Element content	LKSD-2	1989
Lake sediment	CCRMP	Element content	LKSD-3	1989
Lake sediment	CCRMP	Element content	LKSD-4	1989
Petroleum coke	CCRMP	Sulphur	PC-1	1990
Petroleum coke	CCRMP	Sulphur	PC-2	1990
Petroleum coke	CCRMP	Sulphur	PC-3	1990
Stream sediment	CCRMP	Element content	STSD-1	1989
Stream sediment	CCRMP	Element content	STSD-2	1989
Stream sediment	CCRMP	Element content	STSD-3	1989
Stream sediment	CCRMP	Element content	STSD-4	1989
Marine sediment	NRC	Element content	MESS-2	1993
Marine sediment	NRC	Element content	BCSS-1	1981
Marine harbour sediment	NRC	Butyltin	PACS-1	1987
Nearshore seawater	NRC	Element content	CASS-3	1994
Open ocean seawater	NRC	Element content	NASS-4	1992
Estuarine water	NRC	Element content	SLEW-2	1994
Riverine water	NRC	Element content	SLRS-3	1994
Dogfish muscle tissue	NRC	Element content	DORM-2	1994
Dogfish liver tissue	NRC	Element content	DOLT-2	1994
Non-defatted lobster	NRC	Element content	LUTS-1	1989
Ground whole fish	NRC	Dioxins, furans, PCB	CARP-1	1994
Lobster tissue	NRC	Element content	HS-3	1987
Natural marine sediment	NRC	PAH (16)	HS-3	1987
Natural marine sediment	NRC	PAH (16)	HS-4	1987
Natural marine sediment	NRC	PAH (16)	HS-5	1987
Natural marine sediment	NRC	PAH (16)	HS-6	1987
Natural marine material	NRC	PCB	CS-1	1982
Natural marine material	NRC	PCB	HS-1	1982
Natural marine material	NRC	PCB	HS-2	1982
Water	CENAM	Mercury	DMR-2g	
Water	CENAM	Cromium	DMR-3e	
Water	CENAM	Cromium	DMR-3f	
Water	CENAM	Cyanide	DMR-4f	
Water	CENAM	Phenol	DMR-5e	
Water	CENAM	Metals	DMR-8c	
Water	CENAM	Fluorine	DRM-9c	
Water	CENAM	Silver	DRM-11d	
Gasoline	CENAM	Sulphur	DRM-34	

Table 5.5. List of producers

Country	Abbreviations	Institutions
BELGIUM	BCR	Institute for Reference Materials and Measurement, Geel
AUSTRIA	IAEA	International Atomic Energy Agency, Vienna
USA	NIST	National Institute of Standards and Technology, Gaithersburg
GREAT BRITAIN	LGC	Laboratory of the Government Chemist, Teddington
JAPAN	NIES	National Institute for Environmental Studies, Ibaraki
CHINA	NRC	National Research Centre for CRM, Beijing
	RIG	Research Institute of Geology, Beijing
	COCRI	Cereal and Oil Chemistry Research Institute, Beijing
	FDSI	Food Detection Science Institute, Beijing
	IEC	Institute of Environmental Chemistry, Beijing
	SINR	Shanghai Institute of Nuclear Research, Beijing
	IGGE	Institute of Geophysical & Geochemical Exploration, Beijing
	SITT	Shanghai Institute of Testing Technology, Beijing
	BMEMC	Beijing Municipal Environmental Monitoring Centre, Beijing
	EMCHJ	Environmental Monitoring Centre in Heilong Jiang, Beijing
	SMI	Second Marine Institute of National Metrological Bureau, Beijing
SLOVAKIA	PB-ANAL	RN Dr. Stefan Bartha, Kosice
	PPRI	Paper and Pulp Research Institute, Bratislava
NETHERLANDS	WAU	Wageningen Agriculture University, Wageningen
	RIZA	The Netherlands Institute for Inland Water Management and Waste Water Treatment, Lelystad
CANADA	CCRMP	Canadian Certified Reference Materials Program Canmet, Ottawa
	NRC	National Research Council, Ottawa
MEXICO	CENAM	Centro Nacional de Metrologia, El Marques

5.4   References

5.1. Kenda AS, Wade JJ, Ridge D, Poland A (1974) J. Org. Chem. **39**, 931

5.2. Document of the "International Organization for Standardization", ISO GUIDE 30:1992

5.3. Klich H, Walker R (1993) Fresenius J. Anal. Chem. **345**, 104–106

5.4. Guidelines for the production and certification of BCR reference materials (BCR: Community Bureau of Reference) Doc. BCR/01/97, Part A, 15 April 1997

5.5. Guidelines for the production and certification of BAM reference materials, BAM Document, 1997

5.6. Kramer G N, Muntau H, Maier E, Pauwels J
(1998) Fresenius J. Anal. Chem. **360**, 299–303

5.7. van der Veen A M H, Nater D A G (1993) Sample preparation from bulk samples: an overview; Fuel Processing Technology **36**, 1–7

5.8. ISO/TR 11046: Technical Report Soil Quality – Determination of mineral oil content – Infrared spectrometry and gas chromatographic methods. 1st edition 1994-06-01

5.9. ISO guide to the Expression of Uncertainty in Measurement, 1993

5.10. ISO Guide 35, Certification of reference materials – General and statistical principles. 2nd edition, 1989

5.11. CEN-Draft proposal TC 19/WG 22 from 01.11.1993 "Petroleum products and used oils – Determination of polychlorinated biphenyls"

5.12. CR Software for the evaluation of interlaboratory certification study (Version 1996)

5.13. ISO Guide 43, Proficiency testing by interlaboratory comparisons, Second edition 1997

5.14. ISO 5725, Accuracy of Measurement Methods and Mesurement Results, Draft 1996

6 Reference Materials in Clinical and Forensic Toxicological Analysis

Fritz Pragst and Wolf-Rüdiger Külpmann

In clinical as well as forensic toxicology a wide variety of illicit and therapeutic drugs, pesticides or chemicals used in daily life and their metabolites are to be identified and quantitatively determined in human blood, urine or tissues. Preferred methods are immunoassays and chromatographic methods combined with UV or mass spectrometry. Reference materials in this field are equally important for a correct qualitative and quantitative analysis and are available as pure compounds, solutions in water or other solvents, and in serum or urine in liquid or lyophilized state. In forensic investigation of offences under influence of alcohol, standard solutions of ethanol and congener alcohols in water, serum or whole blood are used for calibration and as controls, as well as reference sera for measurement of clinical parameters characterizing chronic alcohol abuse. Reference materials in urine used for calibration and control of immunoassays in detection of illicit and therapeutic drug abuse contain in most cases only one drug, which is representative for a whole substance group with similar structure and effect. Therefore these methods are primarily destined for qualitative assays. Deuterated standards of many drugs have been established as advantageous internal standards for quantitative determination of very low concentrations in human material by gas chromatography/mass spectrometry. Many therapeutic drugs of toxicological importance are available in serum or whole blood as reference materials destined for therapeutic drug monitoring. For most compounds metabolites are still difficult to obtain, and in particular cases, must be gained from voluntary or patient urine samples for comparison of analytical signals. In reference materials for control of drug analysis in hair, homogeneity is a particularly difficult problem. Some commercially offered reference materials are given.

6.1 Problems of Human Toxicological Analysis

In *clinical toxicological analysis* a wide variety of toxic compounds have to be identified and quantitatively determined in human material in emergency cases after inadvertent or deliberate intake (e.g. in accidents or attempted suicides) or in cases of illicit or medical drug abuse [6.1,2]. The majority of analytes are therapeutic and illicit drugs, pesticides, phytotoxins or chemicals of household origin and are non volatile organic substances. Ethanol may be involved in every case. But other volatile substances and some inorganic poisons like CO, alkali cyanides or nitrites are also to be expected. Toxic inorganic elements like As, Hg, Pb or Tl have lost their historically predominant role in acute toxicology and are mainly a subject of occupational medicine, although they should not be disregarded. Usually blood (serum or plasma), urine and in some cases stomach content or gastric lavage fluid are used for analysis. As treatment may be highly dependent on identification and quantitative determination of the compounds involved, the analysis must be carried out as fast as possible.

In *forensic toxicology* the same compounds as mentioned above have to be sought in order to elucidate the cause of death in non-natural fatalities or to prove or exclude an effect of these compounds on the behavior in e.g. traffic, criminal actions or work accidents [6.3,4]. In addition to the sample materials mentioned above in postmortem cases, tissues of liver, kidney, lung or brain are also investigated. For the retrospective detection of e.g. drug abuse, hair samples are examined. A borderline field is the detection of prohibited drugs in doping control [6.5].

A particular problem in clinical as well as forensic investigations is the systematic toxicological analysis, i.e. a general search for any toxic compounds in order to establish or exclude an acute intoxication as cause of death, illness or adverse behavior. In this case as many as 10 000 compounds have to be taken into account, which may have a lethal effect at quite different blood concentrations between 10 µg/l (e.g. the cardiac glycoside digoxin) and 1 g/l (e.g. the analgetic salicylic acid). In drug abuse cases the concentration may be as small as 0.1 µg/l (e.g. the hallucinogenic lysergide, LSD). Usually apart from the mother compound its metabolites are also present in blood and particularly in urine. Furthermore constituents of the matrix, of the normal food or in lethal cases putrefaction products may add to the complex mixture of compounds. In the systematic toxicological analysis the unambiguous identification of the xenobiotics is the main problem, whereas in view of the large biological variability of the dose–effect relationship and with respect to the urgency of the investigation, a very high degree of accuracy in quantitative determination of the identified substances is less important.

This is quite different from some special forensic investigations where, by law or by agreement, limit concentrations or cutoff values are introduced, above which a criminal prosecution may occur. For example in Germany driving a car with an ethanol concentration in full blood above 1.10 mg/g is a criminal offence. In a similar way the presence of illicit drugs in urine is only regarded as positive above certain threshold concentrations (cutoff values). In these cases highly reliable quantitative analytical methods are required.

Over the last few decades the preferred methods used for the detection of toxic compounds in human material have become immunoassays and combined chromatographic–spectroscopic methods like gas chromatography/mass spectrometry (GC/MS), high performance liquid chromatography/UV spectrometry by photodiode array detection (HPLC/DAD) and recently liquid chromatography/mass spectrometry (LC/MS). In chromatography a more or less complicated extraction is usually necessary in order to separate the analyte from the biological matrix and to avoid a superimposition by matrix signals. Substance identification as well as quantitative determination are generally based on the comparison of the analytical responses from the sample and those from the corresponding reference compounds.

Consequently, in clinical and forensic toxicological analysis there is an urgent need for such compounds as highly purified and well characterized reference substances in a stated concentration in water or other solvents, blood, serum, urine or hair. Such materials are commercially available as:

- certified reference material (CRM),
- reference materials (RM),
- pharmaceutical reference materials (PRM),
- conventional laboratory reagent standards.

RM and CRM (for definition see Sect. 2.1) are produced according to ISO Guides 30 and 35. PRM are produced according to corresponding monographs of international or national pharmacopoeia, e.g. the European Pharmacopoeia (EP) or the United States Pharmacopoeia (USP). The production or distribution of high quality and certified standard solutions is mainly carried out by companies such as Sigma, Radian or Promochem.

The intended use of a reference material of higher metrological order, i.e. calibration and trueness assessment, and evaluation of the performance of a measurement procedure, requires a thorough description of the material (according to a CEN draft), among other things:

- warning and safety precautions;
- material itself (e.g. aqueous solution, urine);
- relevant components;
- kind of quantity (e.g. amount of substance concentration, mass concentration);
- issuing authority or manufacturer;
- identification code and lot identification;
- reference material characteristics;
- origin and nature of the material (e.g. inorganic, synthetic, natural);
- matrix (e.g. distilled water, serum);
- relevant components named according to an internationally accepted nomenclature and their concentration (including expanded uncertainty);
- measurement procedure used for assigning values to the reference material and outline of traceability;
- intended use of the reference material and for what purpose it is not applicable;
- required pretreatment of material;
- absence/presence of infectiveness markers;
- pertinent additives and their concentration;

- storage conditions of the unopened container as well as of the reconstituted material;
- details of how to reconstitute the material;
- certificate including items specified in ISO Guide 31.

These recommendations primarily apply to reference materials of higher metrological order. They should, however, be followed as far as applicable and reasonable in the description of reference materials of lower metrological order, which are more often used in clinical and forensic toxicology.

Table 6.1. Commercially available human reference materials for negative control or for dilution of more highly concentrated samples

Material	State	Manufacturer, delivery firm	Remarks
Human serum SRM 909A	Lyophilised	Promochem	Certified for clinical chemistry
Human serum Medidrug Basis-line S	Lyophilised	Medichem	Stability 3 years, after reconstution 30 days.
Lyphocheck drug free human serum	Lyophilised	Biorad	Stability 3 years, after reconstution 10 days
Blank human serum Defibrinated plasma	Frozen	Utak	Stability frozen 12 months, after thawing 12 days
Human urine Medidrug Basis-line U	Lyophilised	Medichem	Stability 3 years, after reconstution 30 days
Lyphocheck urine toxicology negative control	Lyophilised	Biorad	Stability 3 years, after reconstution 10 days
Drug free urine	Frozen liquid	Utak	Stability frozen 9 months, after thawing 30 days

Only a limited number of reference materials in matrix are commercially available. If lacking, it is common practice to prepare the reference material by spiking an adequate matrix. Some commercially available human reference materials used for this purpose and as negative control are given in Table 6.1. It is important that such material be tested and found negative for HIV or hepatitis B virus. During the following preparation steps, the material may be altered, and therefore in many cases does not possess the properties of actual samples [6.6]:

- recalcification of plasma,
- centrifugation,
- filtration,

- stirring,
- freezing/thawing,
- deactivation of serum by heat or chemicals,
- addition of buffer and stabilizers,
- addition of the analyte,
- lyophilization.

Therefore it is sometimes more favorable to use blood or urine freshly sampled from an appropriate volunteer for spiking. Moreover it is recognized that the matrix of some forensic specimens may be "unique" in some way (e.g. putrefied or embalmed) and it may be very difficult or impossible to obtain a similar matrix for preparation of reliable calibrators or controls [6.7]. Hence, very often the method of standard addition may be actually preferred to a conventionally calibrated assay, i.e. the same sample is analyzed before and after addition of known amounts of the analyte.

In contrast to many other fields, the identification and quantitative determination of the majority of toxic compounds from human samples is not carried out in large series, but may even occur only once over a period of years.

External quality assessment schemes for drug screening and therapeutic drug monitoring are performed by several organizations (Table 6.2 [6.8]). The evaluation of quantitative results is performed in different ways in the different schemes. They are assessed e.g. according to the medical requirements of accuracy and in comparison to reference method values (Germany) or with regard to the pertinent consensus mean of the method and stated levels of imprecision (United Kingdom). For qualitative testing the main target is to warrant adequate analytical specificity and sensitivity. The concentrations presented in Table 6.3 are used to decide whether the procedures are sufficiently sensitive.

6.2   Qualitative Analysis

As shown above, a wide variety of reference materials are needed in a clinical or forensic toxicological laboratory. Many of the reference compounds are used in pharmaceutical, food or environmental analysis as well.

With respect to therapeutic drugs many compounds are purchased by manufacturers like Sigma or Promochem. They meet the requirements of the corresponding monographs of the national or international pharmacopoeias. Otherwise the pertinent pharmaceutical manufacturers can help by delivering the substance with a certificate of analysis, in which the identity of the substance is proven by IR, NMR and mass spectrum, and the purity is stated as investigated e.g. by HPLC.

In this context it is important to realise that most of these compounds are not 100% pure. The degree of purity, nature of contaminants or hydratation water stated in the certificate or on the label of the package must be taken into

Table 6.2. Materials for proficiency testing programmes aimed at drugs of abuse[a]

Organisation	Material	Substances involved	"True concentration"[b]
Soc. Française de Toxicologie Analytique	Pooled human urine, spiked, liquid	Amp, Ben, Bup, Can, Coc, LSD, Met, Opi, Pro	Ref. lab. mean
German Society of Clinical Chemistry, Bonn	Pooled human urine, spiked, lyophilized	Amp, Bar, Ben, Can, Coc, LSD, Meq, Met, Opi, Pro, TCA	Spiked
Centre of Behavioral and Forensic Toxicology, Padova, Italy	Human urine, spiked, liquid	Amp, Bar, Ben, Can, Coc, Met, Opi	Ref. lab. mean
Association for Quality Assessment in TDM and Clinical Toxicology, The Hague, Netherlands	Pooled human urine, spiked, liquid	Amp, Ben, Can, Coc, Met, Opi	Spiked or ref. lab mean
Norway National Institute of Forensic Toxicology, Oslo	Pooled human urine, spiked, liquid	Amp, Bar, Ben, Bup, Can, Coc, Met, Opi, Pro	—
Institut Municipal d'Investigatio Medica, Pharmacology Research Unit, Barcelona, Spain	Sterile pooled human samples, spiked or from controlled excretion studies, liquid	Amp, Ben, Bup, Can, Coc, LSD, Met, Opi, Pro	Ref. lab. mean
Dept. Pharmacology, Ther. & Toxicol., VWCM Heath Park, Cardiff, U. K.	Pooled augmented human urine, spiked, lyophilized	Amp, Ben, Bup, Can, Coc, LSD, Met, Opi, Pro	Spiked or overall mean
Gesellschaft für Toxikologische und Forensische Chemie, Institute of Legal Medicine, Heidelberg, Germany	Pooled urine with constituents of human origin and narcotics, lyophilized human serum	5–8 illicit drugs, medical drugs, other poisons	Spiked
		Amp, Can, Coc, Mor	Ref. lab. mean or overall mean

[a] Abbreviations: Amp = amphetamines, Bar = barbiturates, Ben = benzodiazepines, Bup = buprenorphine, Can = cannabinoides, Coc = cocaine and benzoylecgonine, Met = methadone, Mor = morphine, Opi = opiates, Pro = propoxyphene, TCA = tricyclic antidepressants.

[b] Values accepted as "true concentration": ref. lab. mean = mean concentration determined by reference laboratories; spiked = calculated from amount added; overall mean = mean concentration of all participants.

Table 6.3. European Union threshold values for workplace testing according to [6.32]

Drug	Threshold concentration, [µg/l]
Amphetamine group	300
Cannabinoid group	50
Cocaine or metabolites	300
Opiate group	300
Amphetamine	200
Methamphetamine	200
MDMA[a]/MDA[b]/MDE[c]	200
Benzoylecgonine	150
Morphine	200

[a] 3,4-methylenedioxymethamphetamine.
[b] 3,4-methylenedioxyamphetamine.
[c] 3,4-methylenedioxyethylamphetamine.

account. Basic or acidic compounds may be delivered as salts with different counterions. For example, codeine is available as monohydrate, hydrochloride dihydrate, sulfate, dihydrogenphosphate hemihydrate, dihydrogenphosphate sequihydrate, or campher-3-sulfonate, which all have a different content of codeine base.

Furthermore, instructions about the use of the material must be followed carefully. Some compounds must be protected from light, stored at low temperature or protected from moisture. Despite careful storage they may have limited shelf-life. Therefore a periodic monitoring of the identity and purity of the substance is advisable.

Only for very few compounds like anticonvulsant drugs is a certified reference serum commercially available that can also be used for toxicological analyses. In Germany according to the "Guidelines of the Bundesärztekammer" [6.9] the certification of accuracy control sera for carbamazepine, digoxine, phenobarbital, phenytoin, primidone, theophyllin and valproate are performed by reference institutions, which are authorized by the "Bundesärztekammer". Lyophilized bilevel or trilevel therapeutic drug monitoring control sera are also available for other therapeutic drugs (Table 6.4).

Apart from illicit and therapeutic drugs, pesticides are rather frequent in human intoxications. An almost complete list of pesticide agents presently in use is given in the "Pesticide Manual of the British Crop Protection Council" together with a description of chemical properties, toxicology and analytical methods including literature service [6.10]. Reference substances originally prepared for pesticide residue analysis are certified by national and international authorities in most cases according to ISO and are sold by Riedel de Haen (PESTANAL Program), Dr. Ehrensdorfer GmbH or Promochem.

Table 6.4. Commercially available sera for accuracy control in therapeutic drug monitoring.[a] (In most cases one material is intended for control of several drugs and sometimes also other quantities of clinical chemistry)

Drug/matrix (manufacturer/delivery firm)

N-acetylprocainamide/S (AB, CI, DB, RO, SI)

Alprazolam/S (UT)

Amikaine/S (AB, BI, CI, DB, ME, RO, SI, UT)

Amiodarone/S (UT, CS)

Amitriptyline/S (BI, CI, DB, ME, UT)

Amobarbital/S (CI, DB, UT)

Benztropine/B (UT)

Bupropion/B (UT)

Butalbital/S (UT)

Caffeine/S (BI, CI, DB, SI, UT)

Carbamazepine/S (AB, BI, CI, CS, DB, ME, PR, RO, SI, UT)

Carbamazepine-10,11-epoxide/S (CS)

Carisoprodol/S (UT)

Chloramphenicol/S (BI, CI, SI, UT)

Chlordiazepoxide/S (BI, UT)

Chlorpromazine/S (UT)

Clobazam/B (UT)

Clomipramine/S (UT)

Clonazepam/S (BI, CI, DB, SI, UT)

Clozapine/S (UT)

Cortisol/S (AB, BI, CI, DB, RO, SI)

Cyclosporine/S (AB, BI, CI, DB, SI)

Desalkylflurazepam/S (BI, DB, UT)

Desipramine/S (BI, CI, DB, ME, SI, UT)

Desethylamiodarone/S (UT, CS)

Desmethyldoxepin/S (UT)

Diazepam/S (BI, CI, DB, ME, SI, UT)

Digitoxin/S (AB, BI, DB, ME, RO)

Digoxin/S (AB, BE, BI, CI, DB, UT, RO)

Disopyramide/S (AB, BI, CI, DB, UT)

Doxepin/S (BI, ME, UT)

Encainide/S (UT)

Estriol/S (AB, DB, BI)

Ethchlorvynol/S (UT)

Ethinamate/S (UT)

Ethosuximide/S (AB, BI, CS, DB, PR, RO, SI, UT)

Flecainide/S (AB, BI, UT)

5-Flucytocine (UT)

Fluoxethine/S (UT)

Flurazepam/S (UT)

Gabapentine/B (UT)

Gentamicine/S (BE, BI, CI, DB, ME, RO, SI, UT)

Glutethimide/S (UT)

Haloperidol/S (BI, DB, UT)

Ibuprofen/S (UT)

Imipramine/S (AB, BI, CI, DB, ME, SI, UT)

Indomethacine (UT)

Kanamycin/S (AB, CI, SI)

Lidocaine/S (AB, BI, DB, ME, RO, SI,UT)

Lithium/S (BI, CI, DB, ME, PR, SI)

Lamotrigine/B (UT, CS)

Lorazepam/S (UT)

Meprobamate/S (CI, UT)

Methaqualone/S (BI, CI)

Methoclopramide/S (UT)

Methotrexate/S (AB, BI, CI, ME, RO, SI, UT)

Methsuximide (UT)

Methyprylone/S (CI, UT)

Mexilethine/S (UT)

Netilmicin/S (AB, BI)

Norclozapine/S (UT)

Nordiazepam/S (BI, DB, UT)

Nordoxepin/S (BI, ME)

Norfluoxetine/S (UT)

Nortriptyline/S (AB, BI, CI, DB, ME, SI)

Oxazepam/S (BI)

Oxcarbazine/S (CS)

Paracetamol/S (AB, BI, CI, DB, UT, RO)

Paroxetine/B (UT)

Pentobarbital/S (UT)

Phenobarbital/S (AB, BE, BI, CI, CS, DB, PR, RO, SI, UT)

Phenytoin/S (AB, BE, BI, CI, CS, DB, PR, RO, SI, UT)

Table 6.4. (continued)

Drug/matrix (manufacturer/delivery firm)

Primidone/S (AB, BE, BI, CI, CS, DB, PR, RO, SI, UT)
Procaine/S (CI)
Procainamide/S (AB, BI, DB, RO, SI, UT)
Propafenone/S (UT)
Propoxyphene/B (AB, BI, CI, UT)
Propranolol/S (BI)
Quinidine/S (AB, BI, CI, DB, ME, RO, SI, UT)
Risperidone/B (UT)
Salicylate/S (AB, BI, CI, RO, UT)
Secobarbital/S (BI, ME, UT)
Sertraline/B (UT)
Streptomycin/S (AB, CI, SI)
Sulfamethoxazol/S (UT)
Sultiam/S (CS)

Temazepam (UT)
Theophylline/S (AB, BE, BI, CI, CS, DB, RO, SI, UT)
Thioridazine/S (UT)
Thiotixen/S (UT)
Tobramycine/S (AB, BE, BI, CI, DB, ME, RO, SI, UT)
Trazodone/S (UT)
Triazolam/B (UT)
Trimethoprim/S (UT)
Valproic acid/S (AB, BE, BI, CI, CS, DB, PR, RO, SI, UT)
Vancomycin/S (AB, BI, CI, DB, ME, RO, SI, UT)
Venlafaxine/B (UT)
Verapamil/S (UT)

[a] Abbreviations: AB = Abbott, BE = Beckmann, BI = Biorad, CI = Ciba-Corning, CS = ChromSystems, DB = DADE-Behring, ME = Medichem, PR = Promochem, RO = Roche (Behringer/M), SI = Sigma, UT = UTAK Laboratories Inc., B = whole blood, S = serum.

The various chemical constituents of household, garden and home worker articles (solvents, dyes, tensides) are mostly available from the suppliers of fine chemicals and reagents with adequate purity. However, in many cases the composition of the products is not stated (e.g. on the label) and therefore is not known by the analyst. In urgent cases information about the composition may be provided by poison emergency call centers. Sometimes they may be used to identify signals found in the human material in order to exclude or to ascertain its ingestion.

Considering the large number of compounds which may play a role in toxicology, it is impossible to stock them all in the laboratory or to get hold of them within one or two hours. Therefore, at least in emergency cases, an identification must be possible from the reference compounds at hand.

Methods for substance identification under these circumstances in a general unknown analysis or systematic toxicological analysis (STA) were developed on the basis of thin layer chromatographic R_f (retention factor) values in 10 different standardized systems [6.11], head space gas chromatography for volatile poisons [6.12], gas chromatographic retention indices [6.13], capillary gas chromatography with mass spectrometry (GC/MS) [6.14] and high performance liquid chromatography with UV spectrometry by photodiode array detection (HPLC/DAD) [6.15]. Combined chromatographic–

spectrometric methods GC/MS and HPLC/DAD are generally preferred. Computerized library search using special spectra libraries of toxic compounds [6.14,15] enable a substance identification within a relatively short time and with a relatively high reliability. Nevertheless positive results must be confirmed by another independent method.

Large efforts were made by the DFG Commission for Clinical Toxicological Analysis in order to achieve interlaboratory comparability by standardizing the analytical conditions. In thin layer chromatography R_f-reference compounds for acidic, neutral and basic drugs were published for well-defined, stringent conditions with regard to the TLC-plates, solvents, etc. [6.11]. Similarly in GC the system of Kováts indices was introduced [6.12,13]. The reproducibility of the mass spectra after 70 eV electron impact ionization enables the widespread use of the spectra libraries of Pfleger, Maurer and Weber [6.14], NIST [6.16] or Wiley [6.17]. However, the ongoing performance in the development of capillary columns and of HPLC filling materials has up to now prevented a standardization of capillary GC conditions and especially of the HPLC materials in systematic toxicological analysis. At least the use of acidic conditions in HPLC [6.15–19], which provides the stability of the reversed phases and relatively well-reproducible DAD-UV spectra, is widely accepted. Reference compound systems to calibrate retention times in HPLC have been developed [6.20,21].

In systematic toxicological analysis, where an immense number of characteristic data of reference compounds are always needed, the contradiction is particularly striking between long-term constant and standardized analytical conditions with the corresponding certified materials (e.g. HPLC columns) on the one hand and rapid technical progress, demanding change in these conditions and materials, on the other.

6.3   Forensic Ethanol Determination

Methods used for clinical and forensic determination of ethanol are:

- head space gas chromatography,
- enzymatic determination by alcohol dehydrogenase (ADH method),
- photometric determination based on chemical ethanol oxidation by chromate or molybdate (modified WIDMARK method),
- breath alcohol determination.

Special guidelines for blood alcohol determination have been published for calibration, precision control, etc. [6.22], because of the serious consequences which even a small inaccuracy may have for the person investigated in forensic cases. (For example in Germany for a car driver: $\geq$ 0.50 mg/g for a fine, $\geq$ 0.80 mg/g for suspension of driving licence, $\geq$ 1.10 mg/g as a criminal offence; in other offences: 2.0 mg/g as the lower limit of reduced criminal liability and 3.0 mg/g is the lower limit for absence of criminal liability.) For calibration, aqueous standard solutions in the concentration range from 0.0 to 4.0 mg/g are used. Examples for commercially available reference solutions are given in Table 6.5. In some cases concentrations near 1.10 mg/g are more typical.

In forensic cases the results related to whole blood must be given. Because of the different water content in plasma and in blood cells the investigation of serum or hemolytic serum leads to higher alcohol concentrations than does analysis of whole blood. The serum concentration must therefore be divided by 1.20 for calculation of the whole blood concentration. In the case of hemolytic serum the water content of the sample should be determined individually in order to find the pertinent divisor. Furthermore, the vapor pressure of alcohol in head space gas chromatography is higher in full blood or serum than in aqueous samples. In practice, this effect is taken care of by diluting the sample with water (e.g. $1 + 4$, by volume).

Because of these effects, reference serum samples must be used for quality control. The concentration stated for certified materials is measured in reference laboratories. Examples for use of such materials are described in [6.23,24].

As a second scale for the degree of drunkenness, the breath ethanol concentration (mg ethanol/l breath air) was introduced [6.25]. The calibration of instruments is carried out using air saturated with aqueous standard solutions of ethanol at different temperatures. During measurement the actual temperature of the breath air of the person is determined simultaneously by the instrument and the result is adjusted to $34°C$.

Besides ethanol, concentrations in blood of so-called "congener alcohols" (methanol, n-propanol, i-propanol, n-butanol, i-butanol, 2-methyl-butan-1-ol, 3-methyl-butan-1-ol, acetone, butan-2-ol), may be of interest in order to confirm or dismiss someones's claim about the kind of beverages he or she has consumed [6.26]. Such molecules are mainly formed by fermentation of peptides and are present in characteristic concentrations in different beverages. Head space gas chromatography is used for these investigations. Solutions of mixtures of the relevant alcohols and ketones in low, medium and high concentrations in water, serum and whole blood are available as reference materials (Table 6.5).

Ethylglucuronide, a minor metabolite of ethanol, can be determined by GC/MS in order to prove whether ethanol in a blood sample originates from drinking or whether it has been added during or after sampling [6.27]. Furthermore it prolongs the detection time window for alcohol intake because of its longer half-life as compared to ethanol itself. Samples in lyophilised serum reference materials for quality control are available at different concentrations.

Finally, the detection of chronic alcohol abuse is also important from the clinical and forensic points of view. For this purpose, the activity of γ-glutamyltransferase (γ-GT) or the concentration of the carbohydrate defi-

Table 6.5. Commercially available reference materials for determination of ethanol concentration and investigation of alcohol abuse

Material	Matrix/analyte	Concentration [mg/g]	Source	Remarks/application
Alcohol standard solution	H_2O/ethanol	0.5, 0.8, 1.0, 1.3, 1.5, 2.0, 3.0, 4.0	Merck	Stab. 3 years
Alcohol reference materials	H_2O/ethanol	0.25, 0.50, 0.80, 0.90, 1.00, 1.07, 1.20, 1.50, 2.00, 3.00	Promo-chem	Certified or not certified
SRM1828A	H_2O/ethanol	956.29, 1.487, 2.992	Promo-chem	Certified for primary calibration of blood and breath alcohol measurement equipment
Medidrug Eth. W	H_2O/ethanol	0.5, 0.8, 1.0, 1.1, 1.3, 1.5, 2.0, 3.0, 4.0	Medi-chem	Stability 3 years
Medidrug Eth. S and S-plus	Human serum/ ethanol	0.5, 0.8 1.0, 1.1, 1.3, 1.5, 2.0, 3.0, 4.0	Medi-chem	Stability 3 years
Medidrug Eth. S-X	Human serum/ ethanol	X1, X2, X3, not stated, for precision control	Medi-chem	Stability 3 years
Medidrug Eth. RS	Bovine serum/ ethanol	0.8, 1.1, 2.0, 3.0 for quality control	Medi-chem	Stability 3 years
Liquichek serum alcohol control	Human serum/ ethanol	3 levels, for quality control	Biorad	Stability 3 years, after opening 10 days
Medidrug Eth.VB	Human full blood/ethanol	Level 1 (≈ 0.5), Level 2 (≈ 0.8), Level 3 (≈ 1.1)	Medi-chem	Stability 2 years
Medidrug Congener compounds[a] W, S, S-X, VB, VB-X	Water, serum or full blood/ methanol, n-propanol, i-propanol, n-butanol, i-butanol, 2-methyl-butan-1-ol, 3-methyl-butan-1-ol, acetone, butan-2-on, ethanol	3 Levels: e.g. low = (0.5 μg/ml), medium = (1.0 μg/ml) high = (2 μg/ml)	Medi-chem	Stability 1 year For identification of beverage type, e.g. claimed by an accused to have been drunk after an accident.

Table 6.5. (continued)

Material	Matrix/analyte	Concentration [mg/g]	Source	Remarks/application
Multicompo-nent alcohol calibration kit	Acetone, methanol, ethanol, isopropanol	Aqueous solution, 0.5, 1.0 and 4.0 of each	Radian	
Medidrug Ethyl-glucuronide S	human serum lyophil./ethyl-glucuronide	Level 0 (negative); level 1 (low); level 2 (high) (concentration stated)	Medi-chem	Stability 3 years, after reconst. 30 days; for exclusion of exogenous contamination
Medidrug Eth.+γ-GT S	Human serum/ethanol	Concentrations of ethanol and activity of γ-GT stated	Medi-chem	Stability 2 years For control of chronic alcohol abuse
Multiqual-Kontroll-serum, Tri-level	Human serum/clinical parameters	Ethanol, γ-GT and 70 other quantities	Ciba-Corn-ing	Stability 3 years

[a] W= in water, standard
S and S-X= in human serum, certified for quality and precision control
VB and VB-X in human full blood for quality and precision control

cient transferin (CDT) can be determined [6.28], for example. Methods are controlled by reference sera offered by several firms. As a rule these materials are specific to the method and not intended for general use.

6.4   Drugs in Blood and Urine

The detection of illicit and therapeutic drugs with a high abuse potential is one of the most important tasks of clinical and forensic toxicological analysis. Qualitative abuse control is preferentially carried out with urine, since it provides as a rule higher drug or metabolite concentrations and since its analytical properties are more favorable for analysis than those of blood (e.g. lower protein content, higher optical transmission). However, urine concentrations cannot be interpreted with regard to the state of intoxication or behavioural effects. For this purpose the concentrations of the compounds in blood are much more suitable. Nevertheless, blood and urine are the most important materials in human toxicological analysis with somewhat different fields of application.

6.4.1   Detection of Drugs of Abuse in Urine by Immunoassays

Nowadays, drug screening in urine is primarily carried out by use of immunoassays. Reference materials for use as *calibrators* are usually an integral part of the test kit. It should contain the relevant pure component in a simple well-defined matrix mimicking urine.

When immunoassays are used for qualitative testing for drugs of abuse in urine the following constraints have to be taken into account:

- The component used in the calibration solution may differ from the components in the samples to be investigated:

  - In tests for a group of substances only one representative drug is chosen for the calibration solution. Due to the potentially tremendous differences in cross-reactivity, the signal obtained in such an assay must not be used to estimate the amount of drug present in the sample. An estimation can only be performed if the drug is known, its metabolites are not cross-reacting or absent, interfering drugs e.g. from the same group or other groups can be excluded, and calibration was performed by the relevant drug [6.29]. Cross-reactivity may vary with concentration. Examples are the barbiturates (about 20 different compounds), benzodiazepines (about 30 different substances) or the amphetamines including both illicit drugs and safe drugs, such as phentermine or ephedrine.
  - In tests for an individual component, e.g. methadone, the substance used for calibration may be the mother compound, whereas it is mainly the metabolites that are excreted into the urine. Due to differences with regard to cross-reactivity a quantitative estimation is not reliable. Usually the relevant metabolites cannot be used for calibration as they are expensive and not readily available (cf. Sect. 6.5).

- Immunoassays are rather dependent on the matrix. If the tests should give reliable results for the analyte in urine, the matrix of the calibration solution must behave similarly and urine must be used as matrix for the calibration solution.
- Value assignement is mainly based on weighed-in drugs. A traceability chain, including a reference measurement procedure to determine the concentration of the relevant drugs in the calibrator, is usually not available. However, as only qualitative tests are concerned, this aspect is of minor importance.
- It is self-evident that a reference material used as a calibrator cannot serve as a control in the same assay and vice versa.

Reference materials for use as *controls* are obtainable from several manufacturers (Table 6.6). They should contain the relevant compounds in a matrix as similar as possible to urine. In the case of drugs of abuse, controls are primarily used to check sensitivity of the tests and not as in quantitative

Table 6.6. Manufacturers of reference materials for quality control of drugs of abuse in urine

Manufacturer
H. Biermann Diagnostika
Bio-Rad
Biochemical Diagnostics, Inc.
Diagnostic Products Corporation
DMD
High Standard Products Corp.
Medichem
UTAK Lab

assays to assess trueness. The following particularities must be taken into account:

- Drugs are excreted in the urine as hydrophile metabolites. Unfortunately, metabolites are expensive and not readily available. Therefore controls usually consist of drug-free urine spiked with several mother components. At best, in some external quality assessment schemes, urine from addicts is spiked, so that at least test performance with regard to some metabolites can be evaluated.
- The concentration of the drugs should be in a range to demonstrate an adequately low threshold value of the assay. For internal quality control, concentrations 2 or 3 times the practical sensitivity according to Kutter [6.30] are recommended [6.31] as well as a drug free urine for negative control.
- Value assignment is mainly based on the weighed-in mass of the relevant substances. A traceability chain is not usually established. At best, advanced analytical methods are used to check concentrations of the relevant components, although unfortunately without data on expanded uncertainty. As far as qualitative tests are concerned, this aspect is of minor importance.

It may be speculated that assays by "dry chemistry" are more susceptible to changes of matrix than convential (wet) chemistry.

In external quality assessment schemes for workplace testing in the European Union the threshold values given in Table 6.3 have been published [6.32]. An overview of commercially available reference materials for quality control in urinalysis is presented in Table 6.2 [6.8].

6.4.2 Determination of Drugs by GC/MS Using Deuterated Internal Standards and on Deuterated Reference Materials

In general, loss of analyte during sample preparation is the most serious source of bias in toxicological analysis. In order to minimize this error in GC/MS analysis, internal standards in which hydrogen atoms are replaced by deuterium have proved to be ideal. For example, in morphine or cocaine the N-CH$_3$ group is replaced by N-CD$_3$. Due to their almost identical properties, losses during sample preparation affect the analyte to the same extent as the standard. In the selected ion monitoring mode (SIMS) of the mass spectroscopy detectors, measurement can be confined to the typical ions of the analyte and the corresponding deuterated standard. Because of the high resolution of capillary gas chromatography, the specificity of the SIMS-technique and the advantages of deuterated standards, GC/MS has proved to be one of the most selective and sensitive methods in toxicological analysis. Its use is therefore generally recommended for confirmation of immunoassay results.

Some commercially available forensic reference materials with deuterated substances are given in Table 6.7. Of course, the corresponding normal (non-deuterated) reference substances can also be obtained. The standard solutions are delivered in ampoules (Radian, Sigma). As an alternative, some substance standards are purchased in the lyophilized state in flasks and can be dissolved in the appropriate amount of solvent as required (Lipomed). A certificate of analysis informs about the outcome of exhaustive purity checks. Precision is guaranteed by multiple weightings/triplicate analyses and comparison to previous lots, when available, and homogeneity is demonstrated by random analysis of the ampouled standard. Spectroscopic data, melting point and purity determined by HPLC are stated for crytalline substances.

It is a general problem with the rather expensive solutions of deuterated standards that after opening the ampoule and transferring the solution into a flask for repeated withdrawal of aliquots, the organic solvent (e.g. hexane, methanol, acetonitrile or dimethoxyethane) slowly evaporates from the flask during the necessary openings or during storage, since the caps are not completely tight or untighten with temperature decrease in the refrigerator. As a result the concentration of the deuterated standard increases, leading to erroneous (too low) results. This problem is minimized by storage of the standard solutions in Certan® vessels (Fig. 6.1, [6.33]). These have a capillary opening for filling and for withdrawing aliquots using an appropriate syringe. Even after several hours, no solvent loss is observed with the flask open and filled with hexane at room temperature.

In forensic investigations the demands on quality assurance for a GC/MS method with deuterated ions are high [6.34]. At least three ions should be included, which are characteristic for the compound. They should have a mass as high as possible and be different for the deuterated and non-deuterated analyte or its investigated derivative. One of these "diagnostic" ions should be the molecular ion. The calibration curve should be based on the drug free

Table 6.7. Deuterated standards for GC/MS determination of illicit and therapeutic drugs

Compound	Remarks[a]
6-Acetylmorphine-d6, -d3; -d3; -d3; -d3	Ra, AN; Li, ly; Si, AN; Hi, AN
6-acetylcodeine-d3	Li, ly;
6-acetylethylmorphine-d5	Is, AN
Alprazolam-d5	Ra, Me
7-aminoclonazepam-d4	Ra, AN
7-aminoflunitrazepam-d5; -d3	Ra, AN; Li, ly
Amphetamine-d11, -d10, -d8, -d6, -d5; -d6; -d6, -d11; -d5, -d6, -d11	Ra, Me; Si, Me; Is, Me; Hi, Me
Benzoylecgonine-d8, -d3; -d3; -d3; -d3, -d8	Ra, Me; Si, Me; Is, Me; Hi, AN
Buprenorphine-d4	Ra, Me
Butalbital-d5	Ra, Me
γ-butyrolactone-d6	Ra, Me
Caffeine-^{13}C3	Is, Me
Chlorpromazine-d3; -d3	Ra, Me; Hi, Me
Clomipramine-d3	Ra, Me
Clonazepam-d4	Ra, Me
Cocaethylene-d3; -d3,d8	Ra, AN; Hi, Me
Cocaine-d3; -d3; -d3; -d3, -d8	Ra, AN; Si, Me; Is, Me; Hi, Me
Codeine-d6, -d3;-d3; -d3; -d3; -d3, -d6	Ra, Me; Si, Me; Li, ly; Is, Me; Hi, Me
Codeine-^{13}C	Is, Me
Cotinine-d3; -d3	Ra, Me; Is, Me
Desalkylflurazepam-d4	Ra, Me
Desipramin-d3	Ra, Me
Desmethylflunitrazepam-d4	Ra, Me
Desmethylselegilin-d11	Ra, Me
Diazepam-d5; -d3; -d5	Ra, Me; Li, ly; Hi, Me
Diphenhydramine-d3	Ra, Me
Dothiepine-d3	Ra, Me
Doxepine-d3	Ra, Me
Ecgonine-d3	Hi, Me
Ecgonine methyl ester-d3; -d3; -d3	Ra, Me; Si, Me; Hi, Me
EDDP-d3 (Methadone-Metab.)	Ra, Me
Ephedrine-d3; -d3	Ra, Me; Is, Me
Estazolam-d5	Ra, Me
Fenfluoramine-d10	Ra, Me
Fentanyl-d5; -d5	Ra, Me; Hi, Me
Flunitrazepam-d7; -d3	Ra, Me; Li, ly
Haloperidol-d4	Ra, Me
Heroin-d6; -d3; -d9	Ra, AN; Li, ly; Hi, Me
Hydrocodone-d3, -d6; -d3	Ra, Me; Is, Me
Hydromorphone-d3; -d3	Ra, Me; Is, Me
4-hydroxybutyric acid-d6; -d6	Ra, Me; Is, Me
α-hydroxyalprazolam-d5	Ra, Me

Table 6.7. (continued)

Compound	Remarks[a]
2-hydroxyflurazepam-d4	Ra, Me
4-hydroxyphencyclidine-d5	Hi, Me
α-hydroxytriazolam-d4	Ra, Me
Ibogaine-d3	Ra, Me
Imipramine-d3; -d3	Ra, Me; Hi, Me
Lorazepam-d4	Ra, Me
LSD-d3	Ra, AN
Maprotiline-d3; -d3	Ra, Me; Hi, Me
MBDB-d5	Ra, Me
MDA-d5; -d2	Ra, Me; Li, ly
MDE-d5, -d6	Ra, Me
MDMA-d5; -d3, -d5; -d5	Ra, Me; Li, ly; Hi, Me
Melatonin-d7	Ra, Me
Meperidine-d4; -d4	Ra, Me; Is, Me
Mescaline-d9	Ra, AN
Methadone-d9, -d3; -d3	Ra, Me; Is, Me
Methamphetamine d-14, -d11, -d9, -d8, -d6, -d5; -d5, -d9; -d5,-d9,-d14	Ra, Me; Is, Me; Hi, Me
Methaqualone-d7; -d7	Ra, Me; Is, Me
Methohexital-d5	Ra, Me
Methotrimeprazine-d3	Ra, Me
Methylphenidate-d3	Is, Me
Mianserine-d3	Ra, Me
Morphine-d3, -d6; -d3; -d3	Ra, Me; Si, Me; Li, ly
Morphine-3-β-D-glucuronide-d3 (N-methyl-d3); -d3; -d3	Ra, Me; Li, ly; Is, DMF; Hi, Me
Morphine-6-β-D-glucuronide (N-methyl-d3); -d3	Li, ly; Hi, Me
Nicotine-d4	Is, Me
Nifedipine-d6	Hi, Me
Nitrazepam-d5	Ra, Me
Norbuprenorphine-d3; -d9	Ra, Me; Is, Me
Nordiazepam-d5; -d5; -d5	Ra, Me; Is, Me; Hi, Me
Norfentanyl-d5	Ra, AN
Normeperidine-d4; -d4	Ra, Me; Is, Me
Noroxycodone-d3	Ra, Me
Norpropoxyphene-d5	Ra, Me
Nortriptyline-d3	Ra, Me
Oxazepam-d5; -d5	Ra, Me; Hi, Me
Oxycodone-d6, -d3	Ra, Me
Oxymorphone-d3	Ra, Me
Paracetamol-d4	Ra, Me
Pentobarbital-d5	Ra, Me
Phencyclidine-d5; -d5; -d5; -d5,-d10	Ra, Me; Si, Me; Is, Me; Hi, Me
Phenobarbital-d5; -d5	Ra, Me; Is, Me
Prazepam-d5	Ra, Me
Promethazine-d3	Ra, Me
Propoxyphene-d5, -d11; -d7	Ra, Me; Is, Me

Table 6.7. (continued)

Compound	Remarks[a]
Protriptyline-d3	Ra, Me
Pseudoephedrin-d3	Is, Me
Psilocin-d10	Ra, AN
Ritalinic acid-d5	Is, Me
Secobarbital-d3	Ra, Me
Secobarbital-^{13}C	Is, Me
Selegiline-d8	Ra, Me
Tamoxyphen-^{13}C2,^{15}N	Is, Me
Temazepam-d5	Ra, Me
Δ^9-THC-d3; -d3; -d6	Ra, Me; Is, Me; Hi, Me
11-Hydroxy-Δ^9-THC-d3	Ra, Me
11-Nor-9-carboxy-Δ^9-THC-d9, -d3; -d3	Ra, Me; Hi, Me
Triazolam-d4	Ra, Me
Trimipramine-d3	Ra, Me

[a] Abbreviations. Manufacturer: Hi = High Standards Products Corp., Is = Isotec, Ra = Radian, Si = Sigma, Li = Lipomed. Supplied form: AN = acetontrile solution in ampoule, DMF = dimethylformamide in ampoule, Me = methanol solution in ampoule, ly = lyophilized in flask for dissolving in a solvent of choice.

basic solution and five concentrations in the expected range. In the Forensic Toxicology Laboratory Guidelines of the Society of Forensic Toxicologists (SOFT) at least a 3-point calibration is advised [6.7]. The accuracy should be monitored with certified reference material, whereas for precision control uncertified commercial or self-prepared material with the non-deuterated analyte is sufficient. It is interesting that, as a rule, the retention time of deuterated compounds on a methylsilicon capillary column is by 0.01 to 0.05 min shorter than of the non-deuterated compound.

In many cases, in which the position of deuteration is not involved in the mass spectrometric main fragmentation reactions of the compound and the matrix does not contribute to the corresponding m/e peaks, the sample concentration can be calculated approximately from the chromatographic peak area ratio $A_{analyte}/A_{i.st.}$ of a pair of two corresponding ions of the analyte and the deuterated standard and from the concentration of the added standard.

$$c_{analyte} = \frac{A_{analyte}}{A_{i.st.}} \times c_{i.st.} \cdot \tag{6.1}$$

However, if the deuterated group is involved in the main fragmentation the mass spectra of analyte and standard may differ significantly. As an example the mass spectra of methylenedioxyamphetamine (MDA) and the corresponding fivefold deuterated standard d$_5$–MDA after derivatization with pentafluoropropionic anhydride are shown in Fig. 6.2 together with the fragmentation scheme of the standard. In both cases the molecular ion $M^+ = 325$ and 330 is found. The base peaks $m/e = 135$ and 136 are formed by α-cleavage. The peak 162 in the spectrum of MDA originates from the elimination of trifluoropropionamide. In d_5-MDA two smaller peaks $m/e = 166$ and 167 result from this fragmentation, since either $C_2F_5\text{-}CONH_2$ or $C_2F_5\text{-}CONHD$ is eliminated, with a somewhat higher probability for the latter. In this case the peak area ratio A_{167}/A_{162} can be used for quantitation only after calibration, and, in addition, the ratio A_{330}/A_{325} may differ from the concentration ratio because of isotope effects on the fragmentation rates of the molecular ions.

Another problem hampering the direct concentration calculation according to (6.1) is the residual content of non-deuterated substance in the deuterated standard (isotopic purity), which should be stated in the certificate. For example, in a phencyclidine-D_5 standard the isotopic purity D_0/D_5 is stated to be 0.13% from the mass spectroscopic intensity ratio of the molecular peaks 248 and 243. This means that in an analysis with a 100-fold excess of the standard, about 13% of the phencyclidine peak originates from the standard and not from the analyte. Therefore the concentration of the standard should not be chosen too high, in order to avoid false positive results.

For these reasons and particularly in order to exclude matrix effects also with deuterated internal standards, uncritical use of (6.1) should be avoided and a really accurate quantitation carried out via the calibration curve.

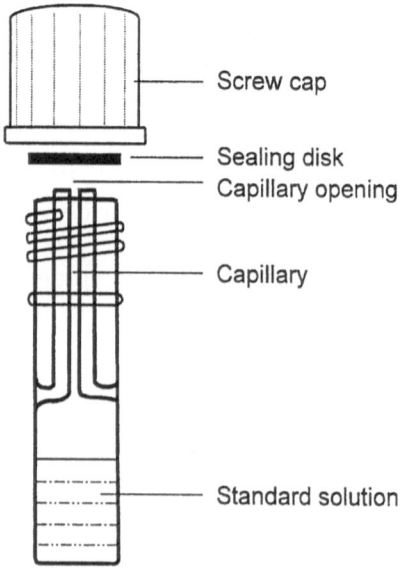

Screw cap

Sealing disk
Capillary opening

Capillary

Standard solution

Fig. 6.1. Certan$^{\circledR}$ vessels for storage of standard solutions with volatile solvents [6.33]. Due to the capillary opening, solvent loss by evaporation is minimized

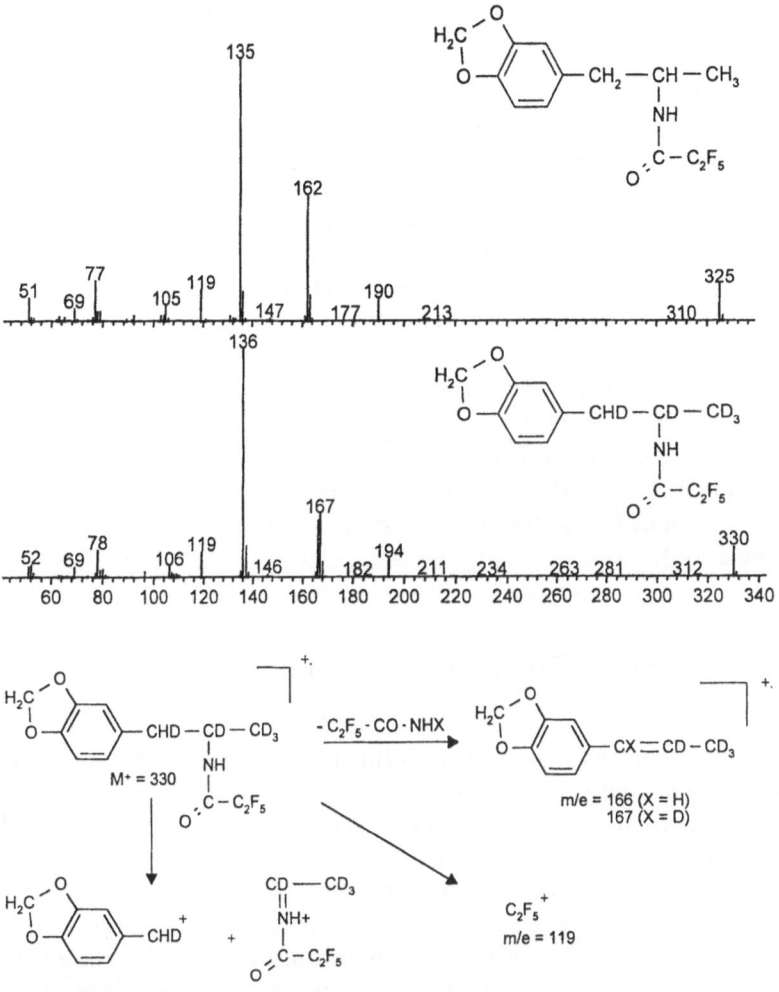

Fig. 6.2. Mass spectra of methylenedioxyamphetamine (MDA) and its pentadeuterated analogue MDA-d_5 after derivatization with pentafluoropropionic anhydride (PFPA), and main fragmentation of PFP-MDA-d5

In contrast to certified reference substances in solution or in lyophilized state, the availability of reference materials containing the drugs in human matrix is rather limited. Some materials for quality control of urine testing in confirmation analysis are given in Table 6.8. Preparation and analytical certification by US National Institute of Standards and Technology (NIST) in cooperation with the College of American Pathologists are described for human urine reference materials of cocaine and benzoylecgonine [6.35], amphetamine and methamphetamine [6.36], morphine and codeine [6.37] and phencyclidine [6.38]. The preparation of three different concentration levels and a blank sample was performed by a commercial supplier from a single urine pool, which was sterile filtered and spiked with the appropriate amounts of drugs. For certification of the concentrations two or more independent and reliable methods were used all based on isotope dilution GC/MS, LC/MS or direct probe MS/MS, flow injection analysis/thermospray mass spectrometry (FIA/MS) with deuterium labeled analogs as internal standards and different procedures of extraction and derivatization. For cocaine, HPLC with cocaethylene as the internal standard was also used. The suitability of the materials was demonstrated by round-robin studies. Of course, the concentrations of the three levels vary from batch to batch.

Similarly the sera materials analyzed during the three round-robins of the GTFCh (Gesellschaft für Forensische und Toxikologische Chemie) every year are prepared in cooperation with the firm Medichem and certified by reference laboratories, and later are available for a certain time as reference materials (Table 6.9). Sometimes the results of the proficiency tests are used for characterization of the material (detected mean value, single and double standard deviation).

6.5   Metabolites in Toxicological Analysis

As a rule an incorporated drug is transformed into several metabolites by more or less complicated enzymatic oxidation, reduction or hydrolysis reactions or by binding to hydrophilic endogenous substances like glucuronic acid, sulfate or glutathione [6.39]. In Fig. 6.3a as an example the HPLC-curve of a blood extract is shown for an intoxication by an overdose of the tranquilizer medazepam. By enzymatic demethylation and hydroxylation five other active benzodiazepines are formed (Fig. 6.3b), of which normedazepam, diazepam and nordiazepam as well as two unidentified hydroxylation products are seen in the chromatogram. For identification and quantitative determination the reference compounds of metabolites are necessary. Since the metabolites are often equally or even more toxic than the mother substance, their concentrations are important for interpretation of the overall toxic effect of the substance intake.

In general, metabolites play an important role in toxicological analysis, for three reasons. Firstly, the mother drug is sometimes already totally metabolized. Secondly their characteristic analytical signals may assist in substance identification, and finally they may contribute to the toxic effect. In most cases metabolites are present in much higher concentration in urine than in blood.

For the majority of medical drugs, metabolism in man as well as in laboratory animals is thoroughly investigated during drug development, and therefore metabolite reference substances may be obtained from the relevant manufacturer. An increasing number of metabolites is also offered by

Table 6.8. Commercially available reference materials for quality control in urine testing for drugs of abuse[a]

Drug	Institution	Stated concentrations [µg/l]	Code
Benzoylecgonine	CAP/Promochem	L 103 M 239 H 510	RM 008
Δ^9-THC-COOH	CAP/Promochem	L 11.7 M 24.1 H 49.6	RM 011
Morphine	CAP/Promochem	L 138 M 293 H 578	RM 019
Codeine	CAP/Promochem	L 134 M 283 H 591	RM 019
Morphin glucuronide	CAP/Promochem	L 210 M 437 H 865	RM 020
Amphetamine	CAP/Promochem	L 311 M 573 H 1137	RM 021
Methamphetamine	CAP/Promochem	L 291 M 555 H 510	RM 021
Cotinine	NIST/FDA/Prom.	L 54 H 488	SRM 8444
Multiple drugs	Promochem	Morphine 309; codeine 288; benzoyl-ecgonine 162; phencyclidine 23.8; Δ^9-THC-COOH 14.8	SRM 1511
Opiates	Medichem	Codeine, codeine-6-glucuronide, morphine, morphine-3- and morphine-6-glucuronide; 2 levels, not stated	Medidrug Opiate U
Multiple drugs	Medichem	Amphetamine, benzoyl-ecgonine, morphine, phency-clidine, oxazepam, secobarbital, Δ^9-THC-COOH; 2 levels, not stated	Medidrug BTM U and BTM U-X
Multiple drugs	Ciba-Corning	Amobarbital, amphetamine, benzoylecgonine, ecgonine, gluthethimide, imipramine, cocaine, meperidine, methadone, methadonemetabolites, methamphetamine, morphine, morphine-3-glucuronide, nortriptyline, oxazepam, phencyclidine, phenmobarbital, propoxyphen, secobarbital, THC, THC-metabolites, quinine; 3 levels according to NIDA; concentrations not stated	DAU – Drogen-mißbrauchs-urinkontrolle

[a]Abbreviations: CAP = College of American Pathologists; NIST = U.S. National Institute of Standards; FDA = U.S. Federal Drug Administration ; L = low; M = medium; H = high.

Table 6.9. Reference materials in human serum or whole blood used in analysis for drug of abuse

Drugs	Manufacturer	Declared concentrations [µg/l]	Code
Amphetamines, cannabinoides, morphine and metab., cocaine and metab.	Medichem	Target value, recommended range and allowable range of concentrations are stated according to GTFCh reference laboratories or investigations in 3 GTFCh round robins every year	Medidrug BTM-A Medidrug BTM-B
Benzoylecgonine, cocaine, codeine, morphine, THC, THC-COOH	Medichem	2 levels for precision control	Medidrug BTM-S
7-aminoflunitrazepam, bromazepam, desmethyldiazepam, diazepam, flunitrazepam, lorazepam, oxazepam	Medichem	2 levels for precision control (S) reference control (plus) with detected values by GTFCh, lyophilized	Medidrug Benzodiazepines-S; S-X; S-plus
Codeine, codeine-6-glucuronide, morphine, morphine-3- and morphine-6-glucuronide; 2 levels, not declared	Medichem	2 levels for precision control (S) reference control S = in serum; VB = in whole blood, lyophilized	Medidrug opiates S Medidrug opiates VB
Amphetamines (10 substances), cocaine and 3 metabolites, opiates (15 substances or metabolites), cannabinoides (9 substances or metabolites), benzodiazepines (31 substances or metrabolites), barbiturates (16 substances)	UTAK	Available on request in blood, serum or plasma	Custom preparations

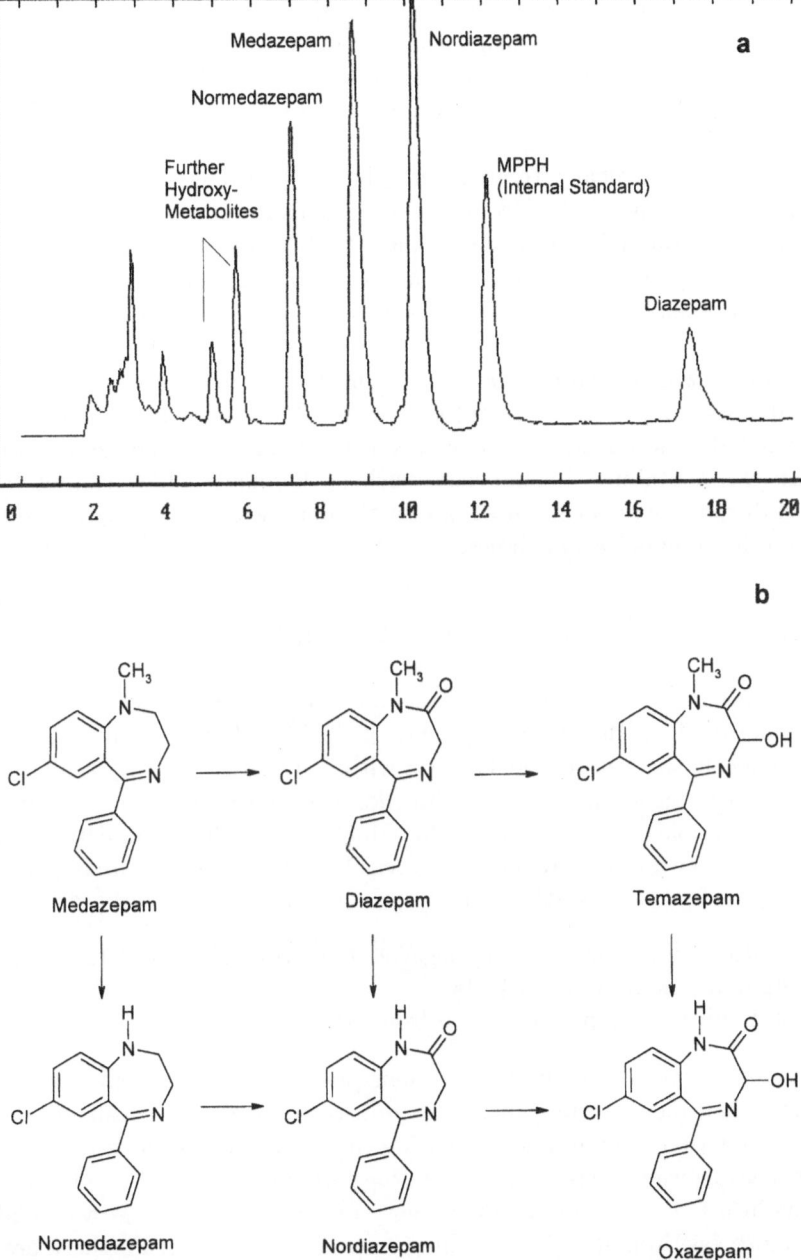

Fig. 6.3. (a) HPLC of an extract of a patient's blood sample collected some hours after intake of a suicidal overdose of medazepam. Apart from the original drug 3 other benzodiazepines and 2 unidentified metabolites are detected as products of biotransformation. (b) Metabolic degradation of medazepam by demethylation and hydroxylation steps

firms like Promochem or Sigma. Furthermore MS and UV spectra of many metabolites are included in toxicological spectra libraries (cf. Sect. 6.2).

Nevertheless, metabolites or reference materials containing metabolites for toxicological analysis are on the whole not readily available. Therefore in serious cases the problem is tackled by using urine or blood samples of patients, who are medically treated with the drug in question, or else the investigator takes a nontoxic dose of the drug. This is also common practice in proficiency testing of doping analysis [6.40,41], where a therapeutic dose of the drug is ingested by volunteers and the cumulative urine is collected and combined in such a way that concentrations of drugs and/or metabolites in the following range are achieved: stimulants 0.5 to 50 µg/ml, narcotics 0.1 to 50 µg/ml, anabolic steroids about 10 ng/ml for the main metabolites, β-blockers 0.5 to 50 µg/ml, diuretics 0.1 to 2 µg/ml. These concentration ranges are expected after administration of banned drugs to or by athletes. Reference urine samples containing illicit drugs (such as heroine), which cannot be produced during volunteer studies, can be obtained with the cooperation of drug addiction rehabilitation clinics.

6.6 Reference Materials in Hair Analysis

Hair analysis has proved itself a useful tool for retrospective detection of drug abuse or poisoning with other compounds [6.42,43]. After the initial use of immunoassays in the last decade, GC/MS with deuterated internal standards has become the most important method for this purpose. A more or less complicated sample preparation procedure, including washing and extraction steps, precedes the analysis. Reference materials with illicit drugs are required particularly in forensic investigations. In principle two approaches are used:

- preparation by soaking of drug-negative hair in a solution of the drugs and its main metabolites [6.44,45];
- use of a hair sample pool collected from addicts [6.46].

The material may be delivered as a hair powder or as hair segments. In comparison to urine or serum, there are additional problems in realizing the properties of actual samples and simulating their behavior during analysis in the best possible way. In spiked hair the drugs are not as firmly bound to the matrix as in a patient's sample after incorporation during hair growth and may be more easy removed in a washing step of the analysis. Furthermore, the amount of metabolites and substances accompanying the drugs is difficult to imitate. On the other hand, if a pool of hair samples of a larger number of addicts is used, it is very important to achieve homogeneity. This is easier for ground hair than for hair segments. But only with hair segments can all steps of the hair analysis be carried out.

Therefore, according to Sachs et al. a pool of hair segments from addicts should be preferred [6.46]. After a first homogenization a homogeneity test is carried out, then the homogenization is continued and checked again (Fig. 6.4). This is continued until the standard deviation of the check stops decreasing. The results found for hair segments by Sachs et al. [6.46] were not as good as for hair powder, but a standard deviation of 0.274 ng/mg at a mean concentration of 1.918 ng/mg 6-acetylmorphine was regarded as acceptable. For each group of illicit drugs (opiates, cocaine, cannabinoides, amphetamines) a separate reference material must be produced. This is necessary in order to avoid, for example, the samples of opiate addicts being diluted by hair of cocaine addicts, which as a rule are negative for opiates. Because of the restricted availability up to now, such material is used only in round robins and for internal quality control.

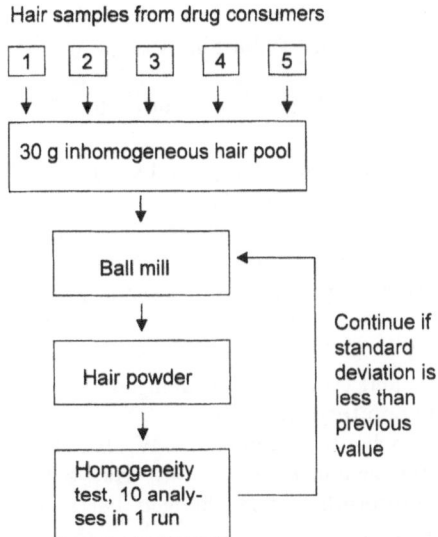

Fig. 6.4. Steps for preparation of a reference material for drug analysis in hair according to Sachs et al. [6.46]

The NIST materials (RM 8448 and RM 8449) commercially available from Promochem were produced from clean drug-free hair samples by soaking several weeks with a dimethylsulfoxide solution containing the drugs until an appropriate concentration had been reached. In order to remove drugs from the hair surface it was merely rinsed with methanol. In the case of hair segments, all hair is taken from one individual, whereas for hair powder, samples from several volunteers were mixed. The reference concentrations were determined with GC/MS–SIM after extraction with 0.1 M HCl at 45°C for 24 h using deuterated standards. Since no further method was used, the materials are uncertified.

Hair powder as a certified reference material has also been prepared for trace-element analysis of human hair [6.47,48]. Such analyses are carried out in laboratories throughout the world for the purpose of assessing the nutritional and toxicological status of a person. The reference material was prepared by washing with water and grinding as much as 6 or 20 kg hair obtained from barber shops. For example, for the material CRM 397 concentrations of As, B, Ca, Cd, Co, Cr, Cu, Fe, Hg, Mg, Mn, Mo, Ni, P, Pb, Se, Sr, Zn were determined in 18 laboratories belonging to 9 European countries. Furthermore, uncertified concentrations of Ag, Ba, Br, Cl, I, La, Mo, P may be stated.

6.7   References

6.1. Ellenhorn M J, Barceloux D G (1988) Medical Toxicology, Diagnosis and Treatment of Human Poisoning. Elsevier Science Publishing Company, New York

6.2. Moeschlin S (1986) Klinik und Therapie der Vergiftungen, 6th edn. Georg Thieme Verlag, Stuttgart

6.3. Baselt R C, Cravey R H (1995) Disposition of Toxic Drugs and Chemicals in Man, 4th edn. Chemical Toxicology Institute, Forster City

6.4. Moffat A C, Jackson J V, Moss M S, Widdop B (1986) Clarke's Isolation and Identification of Drugs, 2nd edn. The Pharmaceutical Press, London

6.5. Clasing D (1992) Doping-verbotene Arzneimittel im Sport. Gustav Fischer, Stuttgart Jena New York

6.6. Will H-G (1998) Clin. Lab **44**, 153–157

6.7. Society of Forensic Toxicologists, Inc. and American Academy of Forensic Sciences (1997) Forensic Toxicology Laboratory Guidelines

6.8. Geldmacher-v.Mallinckrodt M, Hallbach J, Külpmann, W R (1995) Qualitätskontrolle quantitativer Untersuchungen, in: Gibitz HJ, Schütz H (eds) Einfache toxikologische Laboratoriumsuntersuchungen bei akuten Vergiftungen, pp. 83–99, VCH, Weinheim

6.9. Deutsche Bundesärztekammer (1988) Richtlinien zur Qualitätssicherung in medizinischen Laboratorien (RILIBÄK), Lab. med. **12**, 43–60

6.10. Worthing C R, Hance R J (1991) The Pesticide Manual – A World Compendium, 9th edn. British Crop Protection Council, Farnham, Surry

6.11. Deutsche Forschungsgemeinschaft and The International Association of Forensic Toxicologists (1987) Thin-layer chromatographic R_f-values of toxicologically relevant substances on standardized systems, VCH, Weinheim New York

6.12. Deutsche Forschungsgemeinschaft and The International Association of Forensic Toxicologists (1992) Gas chromatographic retention indices of solvents and other volatile substances for use in toxicological analysis, VCH, Weinheim New York

6.13. Deutsche Forschungsgemeinschaft and The International Association of Forensic Toxicologists (1992) Gas chromatographic retention indices of toxicologically relevant substances on packed or capillary columns with dimethylsilicone stationary phases, VCH, Weinheim New York

6.14. Pfleger K, Maurer H H, Weber A (1992) Mass spectral and GC Data of drugs, poisons, pesticides, pollutants and their metabolites, VCH, Weinheim

6.15. Pragst F, Erxleben B-T, Herre S (1997) UV-Spektren toxischer Verbindungen. Photodiodenarray-UV-Spektrenbibliothek von Medikamentwirkstoffen, illegalen Drogen, Pestiziden, Umweltnoxen und anderen Giften. Software und Handbuch, Institut für Gerichtliche Medizin der Humboldt-Universität Berlin

6.16. NIST MS-Library (1993) VCH, Weinheim

6.17. Wiley MS-Library, 6th edn. (1995) John Wiley & Sons

6.18. Turcant A, Premel-Cabic A, Cailleux A, Allain P (1991) Clin. Chem. **37**, 1210–1215

6.19. Nelson J W, Binder S R (1998) Amer Clin. Lab. **12**, 31–34

6.20. Bogusz M, Erkens M (1994) J. Chromatogr. **674**, 97–126

6.21. Hill D W, Kind A J (1994) J. Anal. Toxicol. **18**, 233–242

6.22. Bonte W, Daldrup T, Grüner O, Heifer U, Iffland R, Schütz H, Wehner DH (1997) Richtlinien der Alkoholkommission der Deutschen Gesellschaft für Rechtsmedizin für die Blutalkoholbestimmung für forensische Zwecke, Toxichem + Krimtech **64**, 103

6.23. Aderjan R, Schmitt G (1991) Blutalkohol **28**, 397–404

6.24. Schmitt G, Aderjan R, Schmidt G (1991) Blutalkohol **28**, 325–8

6.25. Mengersen C (1997) PTB-Mitteilungen **107**, 111–114

6.26. W Bonte (1987) Begleitstoffe alkoholischer Getränke. Schmidt-Römhild, Lübeck

6.27. Schmitt G, Aderjan R, Keller T, Wu M (1995) J. Anal. Toxicol. **19**, 91–94

6.28. Heinemann A, Janke D, Püschel K (1998) Blutalkohol **35**, 161–173

6.29. Külpmann W R (1996) Dt. Ärztebl. **93**, A2701–A2702

6.30. Kutter, D (1983) Schnelltests in der klinischen Diagnostik. Urban und Schwarzenberg, München Wien Baltimore

6.31. Geldmacher-v.Mallinckrodt M, Hallbach J, Külpmann, W R (1995) Qualitätskontrolle qualitativer Untersuchungen, in: Gibitz toxikologische Laboratoriumsuntersuchungen bei akuten Vergiftungen, VCH-Verlag, Weinheim

6.32. De la Torre R, Segura J, de Zeeuw R, Williams J (1997) Ann. Clin. Biochem. **34**, 339–344

6.33. Öhme M (1998) LABO Fachzeitschrift für Labortechnik: Heft 3

6.34. Herbold M, Schmidt G (1998) Toxichem + Krimtech **65**, 87–96

6.35. Ellerbe P, Tai SS-C, Christensen RG, Espinoza-Lenuz R, Pauöe RC, Sander LC, Sniegoski LT, Welch MJ, White E (1992) J. Anal. Toxicol. **16**, 158–162

6.36. Ellerbe P, Long T, Welch MJ (1993) J. Anal. Toxicol. **17**, 165–170

6.37. Tai SS-C, Christensen RG, Paule RC, Sander LC and Welch MJ (1994) J. Anal. Toxicol. **18**, 7–12

6.38. Tai SS-C, Christensen RG, Coalkey K, Ellerbe P, Long T, Welch MJ (1996) J. Anal. Toxicol. **20**, 43–49

6.39. K.-H. Beyer (1990) Biotransformation der Arzneimittel, Springer, Berlin Heidelberg

6.40. International Olympic Committee (1995) IOC Medical Code and Explanatory Document, Appendix B: Procedure for Accreditation of Laboratories, Lausanne

6.41. Schänzer W, Donike M (1993) Anal. Chim. Acta **275**, 23–48

6.42. Kintz P (1996) Drug Testing in Hair, CRC Press, Boca Raton

6.43. Kijewski H (1993) Die forensische Bedeutung der Mineralstoffgehalte in menschlichen Kopfhaaren, Schmidt-Römhild, Lübeck

6.44. Pichini S, Altieri I, Pellegrini M, Pacifici R, Zuccaro P (1997) Forensic Sci. Int. **84**, 243–52

6.45. Welch M J, Sniegoski L T, Allgood C C, Habram M (1993) J. Anal. Toxicol. **17**, 389–398

6.46. Sachs H, Kintz P, Staub C (1998) Proceedings of the 50th AAFS Meeting, February 9–14, San Francisco, p. 261

6.47. Quevauviller P, Maier E A, Vercoutere K, Muntau H, Griepink B (1992) Fresenius J. Anal. Chem. **343**, 335–338

6.48. Okamotu K, Morita M, Quan H, Uehiro T, Fuwa K (1985) Clin. Chem. **31**, 1592–1597

7 Use of Reference Materials in Gas Analysis

Bruno Reimann

This chapter aims at a description of the application and proper use of reference materials in gas analysis. It is not a treatise on gas analysis; there are numerous and excellent textbooks available dealing with that subject ([7.1] and [7.2], to name just two). Here, gas analytical techniques will only be mentioned if relevant to the use of gaseous reference materials.

We begin by discussing two terms:

- gas analysis, i.e. chemical anlysis carried out in the gas phase,
- reference materials used in gas analysis.

Gas analysis generally has three goals:

- *identification* of analytes and matrix gases,
- quantitative *determination of analyte content*,
- measurement of *physico-chemical properties* of gases or gas mixtures.

The importance of proper *identification* of analytes and matrix gases has been pointed out in Chap. 2 which dealt with classification of reference materials. Many analytical methods, if not used properly, exhibit cross sensitivities that may induce such severe bias in analytical results that they become useless.[1]

Determination of analyte content is probably the most frequent application of gas analysis. Identification and quantitation of key contaminants are also important goals of analytical work, e.g. contaminants introduced during the manufacturing process of gaseous reference materials (see below).

The most prominent example for the measurement of a *physico-chemical property* of gas mixtures is probably the determination of densities and calorific values of gaseous hydrocarbon mixtures for the natural gas industry. The fact that this can be accomplished by direct combustion in high precision calorimeters as well as by gas chromatographic determination of the composition is an indication of the interdependence of the goals of gas analysis mentioned above.

Certain aspects of the gaseous state and *gaseous reference materials* or calibration gases, as they are often called, have already been dealt with in

---

[1] An example from this laboratory is a GC-NDIR-detector set up for sensitive detection of tetrafluoromethane (CF_4). Due to its construction, this detector is also sensitive to hexafluoroethane (C_2F_6) and other compounds containing C-F-bonds.

Chap. 2. Gaseous reference materials can be pure gases or gas mixtures. The supplied property is in most cases either gas purity or gas mixture composition. It has been mentioned that reference gas mixtures are normally prepared synthetically by mixing measured amounts of pure gases or gas mixtures of known composition. There are only a few examples of gaseous reference materials that are not manufactured synthetically, one being carbon dioxide from natural sources for different isotopic composition; another being the use of certain fractions of liquid hydrocarbon mixtures from oil refinery plants or of natural gas for the calibration of process control equipment. This chapter, however, does not deal with liquefied gases or gas mixtures. It focuses solely on homogenous gases and gas mixtures. Condensation and liquefaction are treated as issues complicating the proper use of gases and gas mixtures that are to be avoided or taken into account explicitly.

Gaseous reference materials are used mainly for *calibration* of analytical instrumentation, i.e., establishing the relationship between instrument response and analyte content, and for *validation* of analytical methods, i.e. the assessment of whether a given method is fit for the intended purpose.

They are also used for the *identification* of analytes and matrix gases in gas mixtures and of trace impurities in high purity gases.

Proficiency testing has a twofold use:

- as a means to establish the composition of gaseous reference materials and at the same time to establish traceability to SI units or to certified reference materials (CRM);
- for direct measurement of the ability of laboratories to carry out correct analytical measurements by participating in interlaboratory comparisons employing stable, well characterized gaseous reference materials.

Amongst the few *non-gaseous reference materials* directly used in gas analysis, standard solutions for titrimetry are mentioned here for the sake of completeness. Since their use in gas analysis is restricted to the analysis of the liquid phase of absorbed analytes, they can contribute only little to the subject and will not be discussed further.

Since there exist very accurate and well described techniques for the static manufacture and dynamic preparation of gas mixtures, and since the gaseous state is well characterized both experimentally and theoretically, application and proper use of gaseous reference materials seems to be rather straightforward. However, care must be taken with respect to handling and transfer and also to issues of material compatibility, and cleaning and treatment of the inner surface of gas cylinders and transfer equipment. In Sect. 7.1, some of the most important pitfalls that can complicate gas analytical measurements are described together with ways to avoid them.

7.1   Particularities of Gases and Gas Mixtures

According to European international transport regulations for dangerous goods [7.3], a gas is defined as follows:

- at $+ 50°C$ the vapour pressure must be larger than 3 bar;
- at $20°C$ and 1013 mbar it must be completely gaseous.

This definition includes both permanent gases and gases liquefied under pressure. However, as mentioned before, this chapter does not describe the use of liquefied gases, since reference materials used in gas analysis are usually completely gaseous. The reason lies in the fact that different components of a mixture of liquefied gases usually exhibit different vapour pressures. The liquid and gas phase of such mixtures have therefore different composition. During withdrawal from either phase, the ratio of liquid to gas volume changes and therefore the overall composition changes as well. Without employing complex, special equipment (for instance a piston cylinder) the composition of liquid gas mixtures withdrawn from pressure vessels of constant volume varies as these are emptied; thus it is difficult and expensive to maintain constant composition with filling level.

7.1.1   Some Properties of Gases and Gas Mixtures

Compared to solids or liquids, gases are volatile, mostly colorless substances of low density whose molecules are far apart from each other. They are handled and stored in closed, mostly heavy systems made from metal and evenly fill the available space where they perform random motions, undergoing non-reactive[2] collisions with other gas molecules and the walls of the vessel, so that clustering does not occur.

At low pressures, diffusion is quite effective and ensures rapid mixing of the components of a gas mixture. At high pressures, gas mixtures have to be homogenized either by rolling gas cylinders or by thermally enforced convection. Once homogenised, the components of a gas mixture do not separate by themselves. However, condensation of one or several components occurs if the temperature of the vessel or transfer system drops below the dew point of the gas mixture.[3]

Pressure, volume and temperature of a gas show a general dependence usually expressed as gas laws or equations of state. Upon changes in temperature, considerable changes in volume or pressure of the gas can occur. Thus, the analysis of gases and gas mixtures differs in some ways from the analysis

---

[2] This describes the ideal case. In real gas mixtures, wall interactions and chemical reactions of components with impurities frequently occur and influence the stability of gaseous reference materials.

[3] Cold mirror dew point meters for the determination of moisture take advantage of this fact.

of liquids or solids. Gas analysis requires special, closed instrumentation and laboratories in which well controlled conditions are maintained with respect to ventilation, ambient temperature and humidity.

7.1.2   Safety Considerations

Gases and gas mixtures are classified

- flammable/pyrophoric,
- oxidising,
- corrosive to material and/or tissue,
- short and long term toxic,
- dangerous for the environment.

Safety considerations have to take this into account in combination with the additional risk of high pressure. The risk of asphyxiation due to lack of oxygen is often overlooked when using otherwise harmless gas mixtures (for instance, a few ppm of carbon dioxide in nitrogen) in closed rooms without proper ventilation. One must also bear in mind that most gases and gas mixtures are colourless and only a few exhibit strong odours that might warn the user.

Thus, when using gaseous reference materials it is strongly recommended to consult material safety data sheets supplied by commercial gas producers.

7.1.3   Diversity of Combinations and Range of Analyte Content

Approximately 120 pure substances fulfill the definition of a gas given above; this number is doubled if vapours of liquids are also taken into account. Combined with the fact that gases and vapours exhibit complete miscibility, the number of possible combinations is virtually infinite. For reasons of chemical incompatibility, however, not all combinations are possible. Furthermore, many gaseous reference materials are required in only a few matrix gases: nitrogen, air, oxygen, argon, helium and hydrogen are the most common. Even so, the number of possible gas mixtures is practically unlimited.

Analyte content may range from sub-ppb to percentages. From the point of view of a manufacturer of gaseous reference materials, this poses extra problems with regard to the dynamic range of analytical equipment, the limits of detection and quantitation of analytical methods, and the purity of matrix gases.

7.1.4   Special Properties of Some Common Mixture Components

This is not intended to be a complete list. It merely reflects complications during production and application of gaseous reference materials frequently encountered by the author and some customers, thus limiting the stability of gas mixtures and their use as reference materials.

Nitrogen Oxide (NO). This reacts quite easily with oxygen. Any residual O_2 content, present either in the matrix gas used for the preparation of NO mixtures or in the cylinder prior to filling, will result in partial or total conversion of NO to NO_2. Furthermore, especially in the case of reference gas mixtures with NO contents lower than 1 ppm, possible back contamination of the cylinder content due to improper purge procedures occurs frequently. Thus, the reference gas mixture can be permanently destroyed. However, since the reaction of NO with O_2 is third order, time scales for conversion can become so large that the effect of the contamination may not be detected immediately (see Table 7.1). It is therefore important to monitor the time constancy of the content of such reference materials.

Table 7.1. Time scale of the reaction of NO with O_2 at 20°C and 150 bar. NO starting concentration is fixed at 200 ppb, O_2 concentration varies as indicated. The third order rate constant was taken from [7.4]. Time periods given are those for homogeneous gas phase reactions. Real systems may exhibit shorter reaction times due to catalytic wall effects

$[NO]_0$/ppm	$[O_2]$/ppm	Time for 99% conversion of NO
0.2	0.1 (stoichiometric amount)	15 280 years
0.2	1	34 years
0.2	10	3 years
0.2	100	112 days
0.2	1000	11 days
0.2	10000	1 day

Diborane (B_2H_6). This is a thermally unstable substance that decomposes slowly into higher boranes and hydrogen. Therefore, according to le Chatelier's principle, diborane mixtures are most stable if hydrogen is used as the matrix gas. In addition, storage temperature should be kept low.

Halogen Halides (HF, HCl and HBr). These are very corrosive towards metals if residual moisture and oxygen are present in gas cylinders, transfer lines or measuring equipment. The analyte content of reference gases containing these components can be lost to a large extent on the way from the cylinder to the sensor if improper equipment or insufficient purge procedures are used.

Carbon Dioxide (CO_2). This gas exhibits strong adsorption on metal surfaces. Purge and conditioning procedures well suited for less adsorptive components may not be sufficient for the use of reference materials containing CO_2.

Sulphur Containing Compounds (e.g. Mercaptanes, Sulfides and Disulfides with Large Organic Groups). These exhibit strong adsorption as well and are prone to decomposition when passed through metal high pressure needle valves. Here, the remedy is to avoid metal surfaces within the transfer system as much as possible and use instead high-strength plastic materials for the construction of transfer lines and pressure reducing capillaries. Possible complications due to long contact times between these components and cylinder walls have to be solved by other means, of course.

Moisture (H_2O). In mixtures with inert gases this exhibits multilayer adsorption on cylinder walls. As the gas mixture is used, the cylinder pressure decreases and adsorbed H_2O is released from the walls at pressures higher than that of other components. The H_2O content in the gas phase increases in consequence. In this case, a low pressure limit should be used for the withdrawal that is higher than that of other components.

7.1.5   Compressed Gas Cylinders

Compressed gas cylinders are subject to strict regulations throughout Europe and, in fact, throughout the world. These regulations are concerned with marking, labelling, inspection, testing, filling and disposition. These issues, however, are outside the scope of this chapter. Relevant information can be found in documents published by the European Industrial Gases Association [7.5] and the US-based Compressed Gas Association [7.6].

High pressure gas cylinders are made of different materials: carbon steel, stainless steel and aluminium alloys. The material of choice depends on the chemical nature and the range of analyte content of the reference material. Inner surfaces may exhibit chemical impurities and metallurgical faults. Since smooth surfaces are less likely to adsorb large amounts of contaminants or material from previous fillings, electropolished stainless steel, although expensive, is probably the material with the best properties. Great care is taken in cleaning and drying the inner surface of gas cylinders, especially if used for reference materials. Each industrial manufacturer disposes of a number of proprietary pretreament and conditioning steps which eventually define the quality and stability of gaseous reference materials. Finally, the material and the construction type of the cylinder valve can be of decisive importance.

Some precautions for handling and storing compressed gas cylinders are:

1. avoid dropping cylinders or striking them violently together;
2. during intra laboratory transport, cylinder valves must be closed and pressure reducers removed, and valve protection caps must be in place;
3. always use suitable transport devices, such as hand trucks;
4. do not lift gas cylinders by the cylinder cap;
5. prior to use, secure gas cylinders to prevent them from falling over, for example by using adjustable chains or cylinder stands;
6. do not allow cylinders to heat above 50°C;

7. always use personal protection such as safety shoes, safety gloves and, especially in connection with corrosive components, safety glasses;

8. check periodically for valve leak tightness;

9. in laboratories, only cylinders in use should be stored, whilst cylinders not in use should be kept outside the laboratory in suitable storage areas or rooms;

10. laboratory personnel must be trained at regular intervals.

One prominent feature in the production of gaseous reference materials is the common use of hydraulic water tests. This, however, is not ideal for the production of high quality reference materials, since residual moisture in gas cylinders has in most cases negative effects on composition and stability. If possible, ultrasonic tests should be used instead of hydraulic water tests. This has the additional advantage that it can greatly reduce pretreatment and conditioning efforts, if the filling history of the cylinder is known.

7.1.6   Methods of Gas Withdrawal from Cylinders

To ensure the proper discharge of compressed gases and gas mixtures from cylinders, pressure reducing devices must be used. Automatic pressure regulators are most commonly used. There are two basic types: single stage and two-stage. Generally, a two-stage regulator delivers a more constant pressure under varying operating conditions. In some cases, the large inner surface and dead volume of pressure regulators is a disadvantage, especially if used with corrosive components or with gas mixtures with low analyte content. In these cases, the use of high pressure needle valves is recommended, where the gas flow has to be controlled manually. Care must be taken, however, not to block the gas flow downstream from the needle valve, in order to avoid dangerous pressure build up within the analytical equipment. The use of flow monitoring devices such as flow meters is strongly recommended.

In order to transfer the gas or the mixture to the analyser without contamination, leak-tight connections and efficient purge procedures such as purging by repeated pressure build-up must be employed. In critical cases, the use of leak detectors may be advantageous. In some cases the additional evacuation of transfer lines and equipment may be considered, especially if pressure reducers with Bourdon spirals are part of the transfer system.

The user must always be aware of the possibility of back contamination by faulty or insufficient purging. It has been mentioned before that the contents of gas mixtures, especially if they contain small amounts of reactive components, may be completely and permanently corrupted if handled improperly. An indication of contamination or insufficient purging is the appearance of the constituents of ambient air, e.g. nitrogen, oxygen, carbon dioxide, and moisture in the analyser.

204   Bruno Reimann

7.1.7   Disposal of Cylinder Contents

During use, the operator has to ensure proper disposal of gases after they have passed through analytical equipment. Non-dangerous gases may be vented into hoods. Dangerous substances have to be disposed of on site. The user is encouraged to ask the gas manufacturer for advice. Cylinders only partially emptied are best shipped back to supplying gas companies with high volume disposal service.

7.1.8   Transport and Storage of Cylinders

International road transport is regulated by ADR [7.3]. If during transport or storage, the cylinder is exposed to low temperatures, partial or total condensation of the content may occur. In this case, the cylinder must be brought up to room temperature for at least 24 hours and homogenised by rolling.

7.2   References

7.1. Smolková-Keulemansová E, Feltl L (1991) Analysis of Substances in the Gas Phase. **28** in Comprehensive Analytical Chemistry, G. Svehla, Ed., Elsevier, Amsterdam
7.2. Hogan, J D, Ed. (1997) Special Gas Analysis, Wiley, New York
7.3. ADR (Accord européen relatif au transport international des marchandises dangereuses par route) (1997) ECE, Genève Other transport regulations: RID (Réglement international concernant le transport des marchandises dangereuses par chemin de fer) for rail transport IATA-RAR (International Air Transport Association – Restricted Articles Regulation) for air transport ADN (Accord européen relatif au transport international des marchandises dangereuses par voie de navigation intérieure) for water transport
7.4. Baulch D L, Drysdale D D, Horne D G, Lloyd A C (1973) Evaluated kinetic data for high temperature reactions, **2**, Butterworths, London
7.5. European Industrial Gases Association (EIGA), Avenue des Arts 3-5, B-1040 Bruxelles
7.6. Van Nostrand R, Ed. (1990) Handbook of Compressed Gases, The Compressed Gas Association, New York

8 The International Network

Harry Klich

8.1 ISO

ISO, International Organization for Standardization, is a worldwide federation of national standards bodies from some 100 countries [8.1]. The mission of ISO is to promote the development of standardization and related activities in the world with a view to facilitating the international exchange of goods and services, and to developing cooperation in the spheres of intellectual, scientific, technological, and economic activity.

ISO's work results in international agreements which are published as International Standards. The scope of ISO covers standardization in all fields except electrical and electronic engineering standards, which are the responsibility of IEC, the International Electrotechnical Commission. Together, ISO and IEC form the specialized system for worldwide standardization – the world's largest non-governmental system for voluntary industrial and technical collaboration at the international level.

The ISO International Standards are developed within technical committees, subcommittees and working groups. They cover topics such as agriculture, building, information processing systems, photography, transport, medical equipment, metrology, computer languages, materials testing, environment, safety, machine tools, containers, nuts and bolts, to mention only a very few. Some 500 international organizations are in liaison with ISO. By 31 December 1997 the work of ISO had resulted in 11 258 International Standards.

8.1.1 ISO/REMCO Objectives

REMCO is ISO's committee on *reference materials*, reporting to the ISO Technical Management Board.

In addition to its basic activities within ISO, the main aim of the committee, since its inception in 1975 at the recommendation of an international seminar on certified reference materials (CRM), is to continue harmonizing CRMs and promoting their use worldwide.

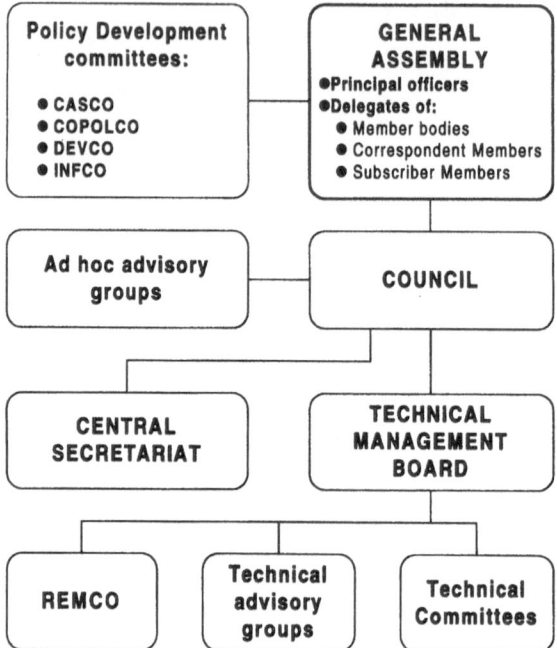

Fig. 8.1. The structure of ISO

8.1.2 Work Programme

REMCO has organized six task groups and distributed the work among them as follows:

I Hierarchy task group

- To consider definitions, categories, levels and classification of reference materials, and recommend actions for REMCO deliberation.
- To provide models for the establishment of traceability of (certified) values in reference materials.
- To contribute to the revision process of VIM, the international vocabulary of basic and general terms in metrology.
- To revise ISO Guide 30:1992, terms and definitions used in connection with reference materials as a result of the revision of VIM.

II Calibration task group

- To study mathematical, including statistical, models of calibration using certified reference materials (CRMs), and to draft appropriate guidance to CRM users for inclusion in ISO guides.

- To develop mathematical models that will assist CRM producers set certified levels when certification data arise from multiple methods, multiple laboratories, method-dependent analyses, or combinations of these sources.
- To collect documentation on calibration of instruments and methods by RMs.
- To revise ISO Guide 33, Uses of certified reference materials, and ISO Guide 35, Certification of Reference Materials – General and Statistical Principles.

III Promotion task group

- To collect and disseminate databases of CRMs already available and in production (including planned CRMs).
- To provide liaison with ISO and IEC technical committees (TCs), international organizations, institutions, agencies and CRM users to identify their needs for CRMs and convey them to producers.
- To inform TCs of CRM availability and encourage mention of CRMs in standards, as appropriate.
- To help organize workshops, seminars and demonstrations, partly in cooperation with DEVCO, to train potential users.
- To study future needs in connection with CRMs and prepare relevant propositions to the ISO Technical Management Board.
- To assist development and dissemination of the index COde of Reference MAterials (COMAR) to advance the promotion of reference materials.

IV Accreditation task group

- To assess the need for the accreditation of RM producers.
- To collect, assess and analyse viewpoints and documentation concerned with the accreditation of RM producers.
- To provide liaison with appropriate national and international organizations concerned with the accreditation of RM producers.
- To coordinate future revisions of ISO Guide 34, Quality System Guidelines for the Production of Reference Materials.
- To draw up rules for accreditation of RM producers.

V Sampling task group

- To collect relevant information with respect to the role of sampling and subsampling in the production and use of CRMs.
- To prepare draft annexes to relevant ISO guides.
- To prepare a bibliography with relevant literature addressing matters related to (sub)sampling of materials.

VI Transportation and Distribution of Reference Materials

- To develop an inventory of difficulties associated with the transportation and distribution of RMs.
- To identify and prioritize problem areas.

8.1.3 Guides Developed by REMCO

ISO Guide 6 (transferred to the ISO Directives Part 2). Methodology for the development of International Standards, Second edition, 1992 Annex B (normative): Mention of reference materials in International Standards.

ISO Guide 30: 1992. Terms and definitions used in connection with reference materials.

ISO Guide 31: 1981. Contents of certificates of reference materials. New revised version available in 1999.

ISO Guide 32: 1997. Calibration in analytical chemistry and use of certified reference materials.

ISO Guide 33: 1989. Uses of certified reference materials. New revised version available in 1999.

ISO Guide 34: 2000. General requirements for the competence of reference material producers. Second edition 2000.

ISO Guide 35: 1989. Certification of reference materials – General and statistical principles.

8.1.4 Other Activities

The booklet "The role of Reference Materials in achieving Quality in Analytical Chemistry" [8.2] gives a brief overview on the application of reference materials, written for those involved in the daily practice of analytical chemistry.

Another booklet with the preliminary title "Introduction to the ISO/ REMCO guides on reference materials" is just under preparation.

There is also the "Worldwide listing of CRM Producers" on the internet. (BAM homepage: http://www.bam.de/a_i/crm/). This database allows the selection of a CRM producer by field of application and country. An example search form is given in Fig. 8.2 and a result screen in Fig. 8.3.

In the next version, specific information on the quality system of the producer will be given, e.g. accreditation or self-declaration.

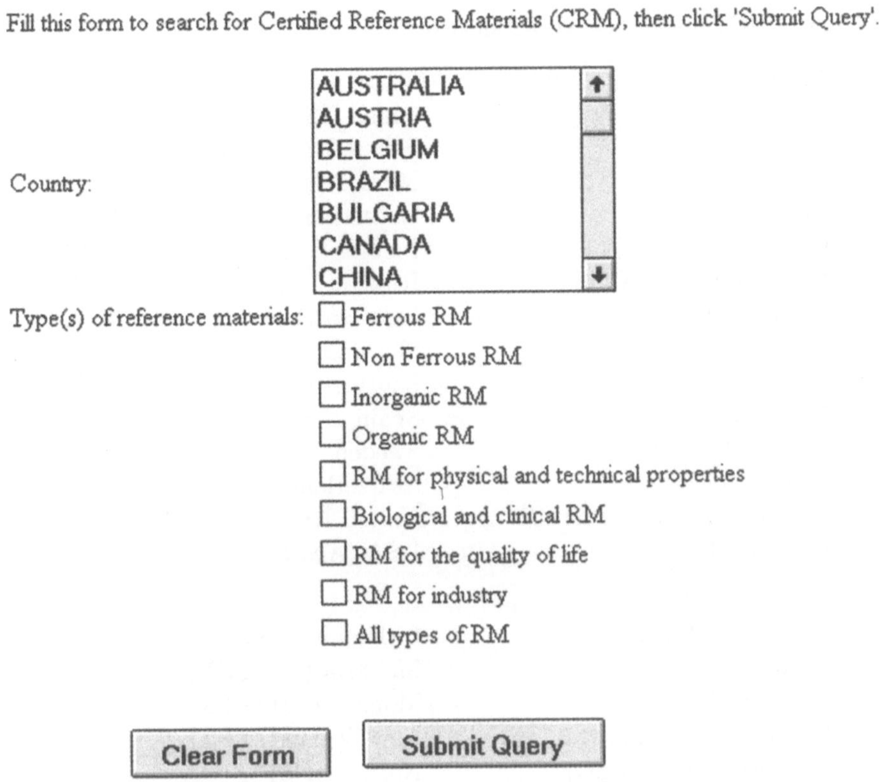

Fill this form to search for Certified Reference Materials (CRM), then click 'Submit Query'.

Country:

AUSTRALIA	↑
AUSTRIA	
BELGIUM	
BRAZIL	
BULGARIA	
CANADA	
CHINA	↓

Type(s) of reference materials: ☐ Ferrous RM

☐ Non Ferrous RM

☐ Inorganic RM

☐ Organic RM

☐ RM for physical and technical properties

☐ Biological and clinical RM

☐ RM for the quality of life

☐ RM for industry

☐ All types of RM

[Clear Form]   [Submit Query]

Fig. 8.2. Search form

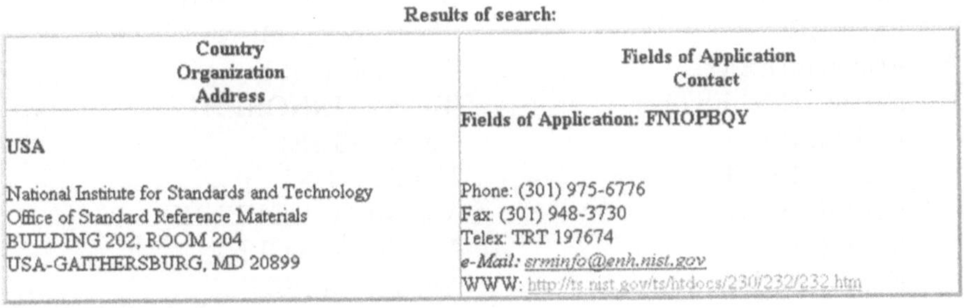

Results of search:

Country Organization Address	Fields of Application Contact
	Fields of Application: FNIOPBQY
USA	
National Institute for Standards and Technology Office of Standard Reference Materials BUILDING 202, ROOM 204 USA-GAITHERSBURG, MD 20899	Phone: (301) 975-6776 Fax: (301) 948-3730 Telex: TRT 197674 e-Mail: srminfo@enh.nist.gov WWW: http://ts.nist.gov/ts/htdocs/230/232/232.htm

Fig. 8.3. Result screen

8.1.5 ISO/REMCO Membership

Participating members (1999)

Australia (SAA)	Japan (ISC)
Belarus (BEST)	Korea, Rep. of (KNIT Q)
Brazil (ABNT)	Mexico (DGN)
Canada (SCC)	Netherlands (NNI)
China (CSBT)	Poland (PKN)
Czech Republic (COSMT)	Russian Federation (GOST R)
Ecuador (INEN)	Slovakia (UNMS)
France (AFNOR)	Slovenia (SMIS)
Germany (DIN)	South Africa (SABS)
Hungary (MSZT)	Spain (AENOR)
India (BIS)	Sweden (SIS)
Indonesia (DSN)	Switzerland (SNV)
Iran, Islamic Rep. of (ISIRI)	United Kingdom (BSI)
Italy (UNI)	USA (ANSI)

Observers (1999)

Argentina (IRAM)	Lithuania (LST)
Barbados (BNSI)	Moldova (MC MDA)
Belgium (IBN)	Mongolia (MNCSM)
Brunei Darussalam (MC BRN)	New Zealand (SNZ)
Colombia (ICONTEC)	Norway (NSF)
Croatia (DZNM)	Portugal (IPQ)
Cuba (NC)	Romania (IRS)
Denmark (DS)	Saudi Arabia (SASO)
Egypt (EOS)	Tanzania (TBS)
Ethiopia (ESA)	Thailand (TISI)
Estonia (MC EST)	Tunisia (INNORPI)
Finland (SFS)	Turkey (TSE)
Greece (ELOT)	Ukraine (DSTU)
Ireland (NSAI)	Venezuela (COVENIN)
Israel (SII)	Viet Nam (TCVN)
Kenya (KEBS)	Yugoslavia (SZS)

8.1.6 International Organizations
in Liaison with REMCO (1999)

BIPM - Bureau international des poids et mesures

EC/IRMM/MT – European Commission – Institute for Reference Materials and Measurements/Measuring and Testing

COWS of WASP – Commission on World Standards of the World Association of Societies of Pathology

ECCLS – European Committee for Clinical Laboratory Standards

IAEA – International Atomic Energy Agency

IEC – International Electrotechnical Commission
IFCC – International Federation of Clinical Chemistry
ILAC – International Laboratory Accreditation Conference
IUPAC – International Union of Pure and Applied Chemistry
OIML – International Organization of Legal Metrology
UNEP-HEM – United Nations Environmental Programme, Harmonization of Environmental Measurement
WHO – World Health Organization
EURACHEM – Association of European Chemical Laboratories

8.2   COMAR: The International Database for CRMs

The need for accurate measurements is becoming ever more important as science and society become more complex and more demanding. In order to achieve such measurements reliably, it is recognised that an analytical laboratory should implement the requirements of a recognised quality assurance system, underpinned by third-party assessment, use validated methodology, participate in proficiency testing schemes where appropriate, use certified reference materials (CRMs) and employ adequately trained analysts.

Of these, the use of CRMs is probably the most important single requirement because CRMs act as the traceability link to the SI international system of measurement. By the application of a CRM whose matrix and analyte composition match as closely as possible that of the samples under test, it is possible for the analyst to ensure that measurements have been properly carried out to the required level of accuracy. This situation has given rise to an ever greater demand for reference materials which are certified for a wide range of compositions and properties.

There are now more than 220 producers of CRMs throughout the world who, between them, produce between 12 000 and 20 000 materials. As a result, it is often difficult to know which is the most appropriate CRM for a given application or, indeed, whether such a material exists at all.

8.2.1   Availability of the COMAR Database

Any chemical analysis and testing laboratory may wish to consider the use of the COMAR database in order to improve the reliability of its measurements or to assist in the development of new methodology. In order to achieve this objective, is is possible to contact the appropriate national coding centre, some of which operate a free advisory service. However, in order to obtain the full benefits available from the database, it is also available for purchase from the coding centres or from the Central Secretariat. In this way it is possible for the user to interrogate the database to a much greater degree than a single enquiry can achieve.

8.2.2 Operating the Database

It is possible to operate the COMAR database using an IBM-compatible personal computer with a minimum space requirement of 20 MB on the hard disk. The programs are written in dBase III and compiled by a ClipperCompiler, running in a DOS window (WINDOWS 3.x, WINDOWS-9x, WINDOWS NT4). Ideally, a printer (matrix or laser) should also be used. The database can be operated in English, French or German. Retrieving information from the database is very user-friendly. Initially the field of application is usually selected, followed by the country of origin, the form of the material and the certified values. In all cases, the user can operate the database in the AND/OR/NOT modes. Certified values can be entered either as specific values or can be set with an upper and lower value range. At the end of a search, the results can be displayed on screen or as a hard copy.

8.2.3 Structure of COMAR

a) Ferrous reference materials

- Pure metal RMs for steel industry analyses.
- Unalloyed steels – euronorm classification (Sect. 8.9).
- Low alloy steels – euronorm classification.
- High alloy steels – euronorm classification.
- Raw materials.
- By-products.
- Cast iron.
- Special alloys (euronorm classification).

b) Non-ferrous reference materials

- Pure metal RMs for non-ferrous metallurgical analyses.
- Li, Be, alkali and alkaline earth metals.
- Al, Mg, Si and alloys.
- Cu, Zn, Pb, Sn, Bi and alloys.
- Ti, V and alloys.
- Ni, Co, Cr and refractory metals.
- Precious metals and alloys.
- Rare earths, Th, U and transuranic elements.
- Raw materials and by-products.
- Other RMs for non-ferrous analyses.

c) Inorganic reference materials

- General inter. Products and reagents (pure).
- Rocks, soils.

- Glasses, refractories, ceramics, mineral fibres.
- Building materials: cements, plasters.
- Fertilizers.
- Inorganic gases and gas mixtures.
- Industrial acids and bases.
- Oxides, salts.
- Other inorganic RMs.

d) Organic reference materials

- Pure organic RMs of general interest.
- Petroleum products and carbon derivatives.
- Synthetic base products, large intermediates.
- Commercial organics: solvents, gases, gas mixtures.
- Plastics and rubbers, organic fibres.
- Paints and varnishes, dyes.
- Cosmetics, surfactants.
- Pesticides and phytocides.
- Fine chemicals.
- Other analytical organic RMs.

e) RMs for the quality of life

- Environment.
- Foodstuffs.
- Consumer products.
- Agriculture (soils, plants).
- Legal controls, criminology.
- Other RMs for quality of life.

f) RMs for physical and technical properties

- RMs with optical properties.
- RMs with mechanical properties.
- RMs with electrical and magnetic properties
- RMs for frequency.
- RMs for radioactivity, isotopes.
- RMs for thermodynamics.
- RMs for physico-chemical properties.
- Other physical and technical properties.

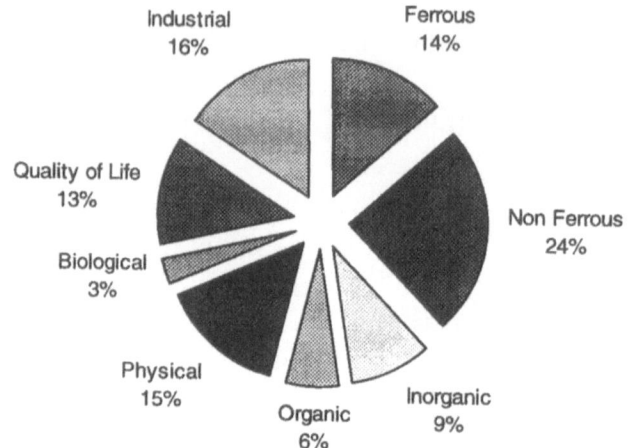

Fig. 8.4. Distribution by fields of application, based on 10 250 CRMs (1999)

g) Biological and clinical RMs

- General medicine.
- Clinical chemistry.
- Pathology and histology.
- Haematology and cytology.
- Immunohaematology, transfusion, transplantation.
- Immunology.
- Parasitology.
- Bacteriology and mycology.
- Virology.
- Other biological and clinical RMs.

h) RMs for industry

- Raw materials and semi-finished products.
- Building, public works.
- Tranportation, communications.
- Electricity, electronics, computer industry.
- Ores, mineral raw materials.
- Measurement and test techniques.
- Fuels.
- Other RMs for industry.

8.3 IAEA

The International Atomic Energy Agency (IAEA) is as an intergovernmental organization and a central participant in a cooperative network of United

Nations organizations, specialized agencies and international organizations enabling the global sharing of information and knowledge in the field of nuclear science.

The Analytical Quality Control Services (AQCS) has provided assistance to laboratories engaged in the assessment of the quality in their work since 1961, especially in the fields of nuclear, environmental and biological materials for major, minor and trace elements, stable isotopes and also organic compounds, using atomic, nuclear and other analytical techniques.

The AQCS programme is involved in the distribution of samples for intercomparison runs and reference materials. There are four main groups of CRMs available:

- nuclear materials and stable isotope products,
- environmental materials,
- biological materials,
- materials of marine origin.

The web site can be found via http://www.iaea.org

8.4   AOAC

Established in 1884, AOAC INTERNATIONAL (formerly the Association of Official Analytical Chemists) is an independent association of scientists in the public and private sectors devoted to promoting method validation and quality measurements in the analytical sciences. The association's primary focus is on coordinating the development and validation of chemical and microbiological analytical methods by expert scientists working in their industrial, academic, and government laboratories worldwide.

In their technical divisions program AOAC established in 1993 the Technical Division on Reference Materials (TDRM). The division recommends policy and criteria to facilitate development and use of reference materials in validation, implementation, and use of AOAC methods, especially the role of reference materials in verification of accuracy.

Their web site can be found via http://www.aoac.org

8.5   European Activities

Community Bureau of Reference (BCR)

Measurement-related activities within the European Commission already have a long history, which began with the launching of the BCR Programme in 1973. This programme started with the certification of RMs in the fields coal and coke, ores and ore concentrates, fertilizers, environment and physical and technical properties.

Today a wide range of BCR Reference Materials is available, covering the most important fields. More information on these materials, e.g. certification reports, can be obtained on the homepage of the Institute for Reference Materials and Measurements (IRMM): http://www.irmm.jrc.be

EC-Standards: Standards, Measurements & Testing Programme

With the advent of the Framework Programme in the 1980s, measurement-related work was redefined in a series of specific RTD programmes, culminating in the programme on Standards, Measurements and Testing (SMT) which ran until the end of 1998. In the 5th Framework Programme (1998–2000), which is composed of four thematic programmes and three horizontal ones, the research on Reference Materials is being continued in the generic programme "Measurements & Testing". Support of inter-laboratory studies and production of certified reference materials (CRMs) is part of the research infrastructures.

The web site can be found at http://www.cordis.lu/smt/src/

EEE-RM

This working group on Reference Materials was set up jointly by EURACHEM, EA and EUROLAB. Their main objectives are:

- to examine the state of the art in this field and promote the correct use of Reference Materials (RMs)
- to identify the essential requirements and information which the producer of RMs must give to the users to provide confidence in the product and to establish comparability in testing
- to provide guidance on proper selection and use of RMs, and action if they are not available
- to define acceptance criteria and validation strategies for RMs (internal and external) RMs, comparability in testing and metrology
- to identify possible ways to promote the dissemination of information on available RMs (e.g., COMAR)

8.6   International and Regional Conferences

BERM

BERM is the International Symposium on Biological and Environmental Reference Materials, which is the only international symposium on reference materials at all. The last meeting, which took place in Antwerp, Belgium, in April 1997 (BERM-7), was attended by more than 200 participants. 99 lessons, 83 posters accompanied by an exhibition of some CRM producers gave an excellent overview on this important field of reference materials.

CERM

CERM is a regional conference on reference materials which will be held for the second time in 1999. Mostly countries from the former eastern block first met in 1996 in the Slovak Republic, and the event was organized by the Slovak Institute of Metrology.

DUREM

DUREM started in 1996 as the first National Workshop on Development and Use of Environmental Reference Materials in New Dehli, India. The second workshop was held in 1999 and gave an overview on the situation in India as well as some border countries, e.g. Bangladesh. The driving force has been the Central Pollution Control Board, which is also deeply involved in REMTAF, the Reference Materials Task Force of India. REMTAF was set up to determine the needs of CRMs in India and to coordinate their production. This programme is sponsored by the Indian government.

8.7 Classification of EURONORM-CRMs

EURONORM-CRMs,[1] prepared under the auspices of ECISS, are grouped into the following categories:

From 001 to 099 – high purity irons and unalloyed steels. No element has a mass content greater than the limit values in the following list:

- silicon, limit value 1.0%
- manganese, limit value 1.5%
- chromium and nickel, limit value for each 0.5%
- cobalt, copper and tungsten, limit value for each 0.3%
- other elements, limit value for each 0.10%
- boron, carbon, phosphorus, lead and sulphur, no limit value.

From 101 to 199 – alloy steels. The content of one or more elements is greater than the limit value in the above list but each does not exceed 5%. The sum of these alloying elements remains under 10%.

From 201 to 299 – highly alloyed steels. The content of one or more element is greater than 5% or the sum of all these alloying elements is at least 10%. Nevertheless the iron content will normally be greater than 50%.

From 301 to 399 – special alloys. The iron content is less than 50%.

From 401 to 499 – pig iron and cast irons.

From 501 to 599 – ferro-alloys.

---

[1] ECCIS - IC1 (Information Circular 1 Final Draft 9/1998).

From 601 to 699 – ores, concentrates, sinters and miscellaneous materials.

From 701 to 799 – additives and refractories.

From 801 to 899 – by-products, such as slags, dusts and similar materials.

The expression "–1" after a CRM number refers to the first issue of a CRM and "–2", "–3", etc. refer to replacement CRMs of generally similar composition, not to a further bottling of the original CRM. Note that this classification is used as a simple and convenient method of differentiating between samples of different types of materials. It has no other objective and does not replace existing European or international standard quality specifications.

The EURONORM-CRMs prepared under the auspices of ECISS do not include coals and cokes because CRMs of these materials are prepared by the Institute for Reference Materials and Measurements (IRMM – formerly BCR).

8.8   References

8.1. Information on REMCO ISBN 92-67-10276-2
8.2. The role of reference materials in achieving quality in analytical chemistry, ISBN 92-67-10255-9

Index

Springer Series in
MATERIALS SCIENCE

Editors: R. Hull   R. M. Osgood, Jr.   H. Sakaki   A. Zunger

* The 2nd edition is available as a textbook with the title: *Laser Processing and Chemistry*